T0262355

Emergency Management

Emergency Management

Edited by **Cathy Hogan**

New York

Published by NY Research Press,
23 West, 55th Street, Suite 816,
New York, NY 10019, USA
www.nyresearchpress.com

Emergency Management
Edited by Cathy Hogan

© 2015 NY Research Press

International Standard Book Number: 978-1-63238-125-5 (Hardback)

Printed in the United States of America.

Contents

Preface

Emergency management is an intricate practice which can save considerable losses during times of calamity if applied efficiently. In the wake of large-scale disasters that we have seen lately, it is obvious that complex and coordinated emergency management systems are needed for efficient and effective relief efforts. Such management systems can only be created by involving various scientists and practitioners from different areas. This book caters to several issues like the impact of human nature, advancement of hardware and software architectures, cyber security concerns, dynamic methods of guiding evacuees and routing vehicles, supply allocation, vehicle routing difficulties in preparing for and responding to large scale emergencies. Researchers, engineers and students of all academic fields would find this book useful. It specially caters to those in the areas of operations of research, human factor and computers. It is also a useful reference for practitioners of emergency management strategies.

The information contained in this book is the result of intensive hard work done by researchers in this field. All due efforts have been made to make this book serve as a complete guiding source for students and researchers. The topics in this book have been comprehensively explained to help readers understand the growing trends in the field.

I would like to thank the entire group of writers who made sincere efforts in this book and my family who supported me in my efforts of working on this book. I take this opportunity to thank all those who have been a guiding force throughout my life.

Editor

Cyber Security Concerns
for Emergency Management

Jessie J. Walker

University of Arkansas at Pine Bluff/Computer Science Unit
USA

1. Introduction

Cyber security has become a matter of national, economic, and societal importance. Present-day attacks on the nation's computer systems do not simply damage an isolated machine or disrupt a single enterprise system. Instead, modern attacks target infrastructure that is integral to the economy, national defense, and daily life. Computer networks have joined food, water, transportation, and energy as critical resources for the functioning of the national economy. When one of these key cyberinfrastructure systems is attacked, the same consequences exist for a natural disaster or terrorist attack. National or local resources must be deployed. Decisions are made to determine where to deploy resources. The question is who makes these decisions? The data required to make and monitor the decisions, and the location of available knowledge to drive them may sometimes be unknown, unavailable, or both.

Indeed, computer networks are the "central nervous system" of our national infrastructure. We are faced with the difficult task of securing our critical cyberinfrastructure from foreign and domestic attacks. In addition, the backbone of emergency management (EM) is a robust cyberinfrastructure. These systems enable emergency management agencies to implement comprehensive approaches to natural disasters, terrorist attacks, and law enforcement issues. There is a general lack of understanding about how to describe and assess the complex and dynamic nature of emergency management tasks in relation to cyber security concerns. Another issue is knowledge integration and how it helps managers improve emergency management task performance. Ever since the first computer virus traversed the Internet, it has been apparent that attacks can spread rapidly. Just as society has benefited from the nearly infinite connections of devices and people through the US cyberinfrastructure, so have malicious parties with the intent of taking advantage of this connectivity to launch destructive attacks.

Surprisingly, very few studies have attempted to tap into the vast knowledge-base of cyber security and emergency management to discover new and relevant theoretical models addressing the two areas. There is also a lack of theories and tools that organizations can use to improve EM success that relate to handling cyber security through effectively managing task complexity and knowledge integration.

This chapter will explore how cyber security concerns related to the uncertainty of emergency management tasks can be addressed for secure EM. In addition, we examine how cyber situational awareness can exploit the mediating role of knowledge sharing and integration to enhance EM tasks.

2. Emergency management community of practice

Friedman and Wyman (2006) theorized that technology has leveled or "flattened" the global playing field that once existed. This flattening has happened as a result of what they call the "triple convergence" of platform, process and people. When an innovative platform takes hold, processes that use the platform must change. This is especially true in contemporary EM communities. If the right people are available, trained, adept and able to adopt technology and process paradigms, they become the third prong of the triple convergence. EM has witnessed this transformation.

In 2007 in both process and technology utilization, Dr. Wayne Blanchard of the Federal Emergency Management Agency (FEMA)'s Emergency Management Higher Education Project (FMHEP) developed the EM community's strongest set of guiding principles for EM, all of which relied on cyberinfrastructure resources (Abbott, Hetzel, & American Bar Association. Section of State and Local Government Law., 2010; Lansford, 2010; LearningExpress (Organization), 2010). These principles are the governing rules that direct EM activities within each EM tasks directly. They include:

1. Develop comprehensive plans which require all emergency managers to take into account all possible hazards, EM tasks, stakeholders, and anticipate all possible impacts to relevant communities;
2. Develop progressive plans that anticipate future emergencies, disasters and develop preventive, preparatory measures to build disaster-proof and elastic communities that are capable of withstanding any type of disaster or emergency;
3. Develop risk-driven models for emergencies and disasters using well accepted EM principles to assign priorities, personnel, and resources;
4. Develop integrated plans to ensure true uniformity among all segments of the EM community, outside organizations and civilian populations;
5. Develop collaborative plans which create true communities of practices;
6. Develop coordination plans which synchronize the activities among community members;
7. Develop and implement flexible plans that can change with the demands of the environment;
8. Develop a professional community, which integrates technology and science in all segments of EM including communal values, and ethics systems.

All of these activities have created within EM a community of practice. Wenger, McDermott, and Snyder (Wenger, McDermott, & Snyder, 2002) define communities of practice as "groups of people who share a concern, a set of problems, or a passion about a topic, and who deepen their knowledge and expertise in this area by interacting on an ongoing basis." According to these authors, communities of practice operate as "social learning systems" where practitioners connect to solve problems, share ideas, set standards, build tools, and develop relationships with peers and stakeholders. Because they are inherently boundary-crossing entities, communities of practice are a particularly appropriate structural model for cross-agency and cross-sector collaborations within EM. The community of practice in EM is now expanding somewhat to incorporate cyber security awareness at all levels (Elmagarmid, Samuel, & Ouzzani, 2008).

Although, the EM community has had difficulty, in defining cyber security awareness in terms of EM governing tasks, this challenge is derived from the peculiar aspects of the field of cyber security. The universe of cyber security is an artificially constructed abstraction that is only weakly tied to physical systems. Therefore, there are few a priori constraints on either the attackers or the defenders. Also, one of the most significant challenges in defining cyber security within the context of EM, is the fact that most of the threats associated with cyber security are dynamic in that the nature and agenda of adversaries is continually changing. In addition, the type of attacks encountered evolves over time, partly in response to defensive actions. Cyber security awareness within EM requires understanding of technology concepts, but also shares aspects of many other disciplines such as epidemiology, economics, and social science. All of these analogies are helpful in providing EM cyber security awareness direction for those within the community (Forrest, Hofmeyr, & Somayaji, 1997; Jennex, 2008).

A recent example of an organization attempting to integrate in cyber security awareness into their EM structure is the California Emergency Management Agency (Cal EMA), which has developed a statewide approach which implements cyber security awareness at all levels of the state's EM plans. The new approach places cyber security activities and concerns alongside other disasters that could possibly impact the state's citizens and infrastructure. The plans include efforts to consolidate its cyberinfrastructure resources to secure data for more than 150 agencies. Although, these efforts are not the result forward-thinking EM personnel, but rather the result of the state experiencing thousands of security breaches in 2010, which were documented by the state's technology staff. As a result of these activities the state, began to see the importance of cyber security in the context of its EM needs (Collins, 2011).

2.1 Technology-driven emergency management

EM can be defined as a unique set of tasks in which, individuals, organizations, governments and nations attempt to bring order to chaos. Emergencies by their very definition are chaotic events brought on by unforeseen and unpredictable circumstances (Bhavanishankar, Subramaniam, Kumar, & Dugar, 2009; Chen, Sharman, Rao, & Upadhyaya, 2008; Mendon, Jefferson, & Harrald, 2007). These events share a unique set of characteristics, which can be identified by the set of associated tasks and the knowledge, which defines the tasks. Davenport and Prusak (Davenport & Pruask, 1998) define knowledge as an evolving set of data that is a mixture of framed experience, values, contextual information and insights defined by experiences for evaluating and incorporating new experiences and data. Knowledge management is at the core of effective EM. The key to knowledge management within EM is, who possesses the knowledge, where is it located and how to find it. Therefore, a significant portion of EM is how to integrate knowledge management and task behavior. According Murphy and Jennex (Murphy & Jennex, 2006) knowledge management within EM is a practice of selectively applying knowledge from past experiences of decision makers to the current and future activities with the purpose of improving individual or organizational effectiveness in terms of the required EM tasks. Knowledge management and dissemination for modern EM tasks are linked directly and indirectly by cyberinfrastructure structures which, consists of computer systems, data and information management, advanced instruments, visualization environments, and cyberspace all linked together by software and complex networks (Elmagarmid et al., 2008; Feng & Lee, 2010; Hong & Lindu, 2009). As a result, cyberinfrastructure enables storage and transfer of massive amounts of knowledge to enable planning, resource allocation, personnel deployment, and coordination of emergency situations (Becerra-Fernandez et al., 2008).

Although, most EM focused organizations possess a significant cyber security situational awareness deficit and how it impacts their reliance on cyberinfrastructure resources. These organization's failures in this arena is evident by recent national and international events. For example, the most recent failure of such systems which hampered effective EM included the attack on 9/11 in which law enforcement/rescue agencies were unable to communicate; Hurricane Katrina in which information coordination was limited or nonexistent; and the recent earthquake in Haiti in which the entire country went totally silent, which made EM almost impossible (Asimakopoulou & Bessis, 2010; Chandler & BCP Media., 2005; Hart, Rudman, Flynn, & Council on Foreign Relations. Independent Task Force on Homeland Security Imperatives., 2002).

Recent events on the international stage demonstrate a similar lack of cyber security situational awareness with respect to cyberinfrastructure resources. In January 2009, the Ministry of Defense in the United Kingdom reported that for two-weeks it did not have access to computers systems within the Royal Navy because of a malware attack which had left the system inaccessible to its personnel. During the same period in the United Kingdom, several hospitals suffered a similar attack, and a month later in February, London hospitals lost all network connectivity due to malware infections that occurred at the end of 2008. At the same time in the U.S., the municipal court system in Houston, TX was infected in a similar manner resulting in a suspension of court proceedings and forcing local police officers to suspend arresting individuals for minor offenses (Saurabh Amin, Litrico, Sastry, & Bayen, 2010; Bayer, Kirda, & Kruegel, 2010; Maughan, 2010; Neumann, 2010).

These examples clearly present evidence that cyber security is now critical to the survival of modern society (Hansen & Nissenbaum, 2009). Clearly, cyberinfrastructure is the infrastructure on which modern homeland security activities depend but security attributes of such resources is often vague or un-measurable. For example, small changes at the bit level in data communication systems can have significant and profound implications that are often poorly understood by the general public, communities that depend on such resources. Cyber security approaches have seen very limited success and have become an arms race with adversaries around the globe. Although, not all communities have been participants in this arms race, but have rather sat on the sideline, and played the role of a victim to evolution. EM as a discipline lacks the fundamental concepts, principles or tools to reliably predict or even measure cyber-security as task components of its activities. It is currently difficult to determine the qualitative impact of evolving cyber security concerns (i.e. more secure now or less secure?) much less quantify the improvement on some specific scale within the domain of EM.

The question for EM organizations is how will they handle cyber security situational awareness within the context of the cyberinfrastructure resources they depend on and how will they develop cyber security abstraction models that exploit the knowledge and experience of sophisticated members of their community as well as provide a framework for discussion of cyber security issues.

2.2 Deterrence and emergency management

Deterrence has proven to be a reliable strategy for ensuring peace with nations, and has been the backbone of international relations since the Cold War of 1946 to 1991 made it an essential element of peace. Deterrence at its core can be defined as preventing an adversary from taking any threatening offensive actions by inducting a set of predefined counter attacks that will convince them they have nothing to gain by the proposed set of actions. However, as the world has changed so have the numerous threats, and the deterrence policies of the past that only countered physical courses of action are no longer as vital to national security as they once were. This is especially true within the sphere of EM, no existing EM plan with in the US contains any type of deterrence policies (Moteff, 2004; Watts, 2003).

As the US continue to infuse society/EM with the world of cyberspace it is essential that we evolve our methods of deterrence and formulate credible threat models that will govern the activities within this new domain. As stated above the US's critical cyberinfrastructure which is the linchpin of EM activities within the US is attacked daily not only from foreign threats but domestic terrorism. The main reason that we find ourselves so vulnerable to such breaches of trust is because we have yet to clearly voice the viable repercussions for those that so choose to impede upon our EM activities beyond standard law enforcement. This highlights a major lack of communication on a local, international scale, and it is upon this lack of consensus that organizations find themselves able to freely commit cyber attacks that can greatly impact EM issues (Harknett, Callaghan, & Kauffman, 2010).

The problem of attribution has plagued cyber security law enforcement ever since the Internet became an accessible form of communication. It prevents victims of cybercrimes from justly placing blame where it should, and thus strips states of the mere ability to even make credible threats of retribution. In essence, when one fights a cyber security attack they are fighting a ghost, and this realization has the power to discredit any proposed counter attacks before they can even be formulated. Knowing this, we've found that when we talk about deterring cyber-attacks in the modern age it is pivotal that we put more emphasis on the aspects of resilience and denial than that of retaliation.

Retaliation can only effectively deter one target group of hackers, and even then its effectiveness is measured based upon the amount of communication and cooperation we receive from international entities. Better security protocols on the other hand can prove their capabilities as soon as they are put into place, and can even abolish the need for threats of retaliation if they are truly impenetrable.

3. Cyber aware emergency management

According to Grant et al (Grant, Venter, & Eloff, 2007), an intrusion within the context of a computer or network system is an act of wrongfully entering, or seizing or taking control of the property of another for malicious purposes. No computer system within our modern society is an island, as a result of an interconnected cyberinfrastructure world where most systems exist in a vast evolving infrastructure of computer network systems. These systems have brought too much of the world vast amounts of information, access to communities once remote and resources that were the stuff of science fiction. These systems have also become an integral part of our modern civilization just as oil was the defining force of the nineteenth century, computer networks have become the defining force for national defense in the twenty-first century. But just as the shipping lanes brought the plague on the back of rats to thirteenth century Europe, these new cyber resources have brought new threats to the shores of the world.

As EM has evolved over the last decade so has the notion of knowledge within the context of cyber security awareness, according to Becerra-Fernandez et al (Becerra-Fernandez, Xia, Gudi, & Rocha, 2007) knowledge at the core of EM can be divided into three specific knowledge types:

Context-Specific Knowledge:

Context-specific knowledge, which is defined as a type of knowledge that is temporal in nature centered around a particular set of circumstances.

Technology-Specific Knowledge:

Technology-specific knowledge is centered on a particular technical toolset, which is comprised of rules used to solve a particular problem.

Context/Technology-Specific Knowledge:

Context/technology-specific knowledge, which is a hybrid knowledge that combines a rich set of contextual knowledge while at the same time possessing a significant technical specificity.

This hybrid knowledge represents a tangent point between EM knowledge, tasks, and cyber security situational awareness. Cyber situational awareness is an emerging aspect of cyber security, in which organizations are aware of all cyber assets they are connected to or depend on (Ke, Ming-Tian, & Wen-Yong, 2009; Kellerman, 2010). Emergency management cyber situational awareness requires individuals, organizations to understand how the resources, events, information, individuals actions impact EM tasks both in the near term and future.

Since all EM tasks fall within the following categories: mitigation, preparedness, response and recovery (Dudenhoeffer et al., 2007), each is impacted by cyber situational awareness as listed in figure 1.

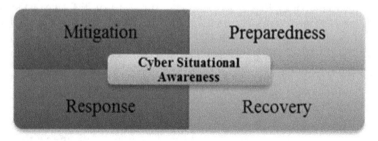

Fig. 1. Cyber Situational Awareness for Emergency Management.

For example, most modern EM tasks needs are met using cyberinfrastructure resources known as Emergency Management Information Systems (EMIS), which support all required tasks of EM. These systems are designed to support interoperability between all required tasks for EM at all segments of the EM community including governmental organizations and civilian populations (Desourdis, 2009; Hong & Lindu, 2009; Moore, 2010). EMIS supports mitigation activities by providing EM personnel the ability to predict, model, and categorize risks using software tools such as geographical information systems (GIS). EMIS allows EM personnel the ability to develop preparedness plans for many types of emergencies modeled on different types of EM scenarios using computer-generated analysis. EMIS provides EM the vast knowledge available in cyberinfrastructure; these services include resource tracking, personnel management, developing and implementing response contingency plans. One of the most significant, support services provided by EMIS to the EM community is the quantification of the true cost of emergencies. Another service includes the development of a uniform community in which remote sensors are connected to provide valuable information to EM personnel to implement future EM mitigation activities (Cohen, 2009). Mitigation tasks are unique within EM, since they are designed to reduce or eliminate risk. Mitigation tasks are either structural or non-structural. Structural tasks typically use technological components in their implementation such as remote wireless sensors or monitoring stations. Non-structural tasks normally include planning techniques such as governmental legislation for land management and development. Mitigation tasks can have the greatest impact on EM since they are designed to prevent emergencies and disasters, according to Rocha et al (Rocha, Becerra-Fernandez, Xia, & Gudi, 2009).

These tasks in an uncertain emergency environment allow planners to close the knowledge gap between specialists in the field and the general population. For example, these tasks and resources are rarely considered when examining cyber security concerns for systems, which are integral to early warning disaster systems and civilian population planning within emergency events such as supervisory control and data acquisition (SCADA) systems (Samia Amin & Goldstein, 2008). Individuals and governmental organizations to ensure resources, personnel, undertake preparedness tasks in the context of EM and infrastructure is ready for any type of emergency that may occur within their boundaries. One of the most important components of preparedness is the development of a communication plan that can be implemented within natural disasters or emergencies. The most recent example of such a failure of preparedness that relates to cyberinfrastructure systems was Hurricane Rita in 2005, one of the most intensive Atlantic hurricanes ever recorded resulting in over 11.3 billion in damage (Davis, Advisory Panel to Assess Domestic Response Capabilities for Terrorism Involving Weapons of Mass Destruction (U.S.), United States. Dept. of Defense. Office of the Secretary of Defense., & National Defense Research Institute (U.S.), 2006; Morris, 2009).

The hurricane deaths were the most severe in open unprotected locations such as roadways, because of poorly executed evacuation plans within the affected regions. Many people were trapped on roadways and interstates because of the lack of communication services (i.e., a failure of computer and data networks), inability of EM personnel to communicate important evacuation instructions on roadways and through standard communication channels. Many EM organizations have only recently started to develop preparedness plans that include cyberinfrastructure services as a component of their overall structure (*Disaster planning and relief. Part 2,* 2010; United States. Federal Emergency Management Agency., United States. Federal Emergency Management Agency. Community Preparedness Division., & Citizen Corps (USA Freedom Corps), 2009). Response is the linchpin of EM. It ensures that the necessary personnel and resources are mobilized when emergencies do occur. These responders include firefighters, police officers, ambulance crews, and in some situations, the National Guard.

These services personnel rely on communication tools, data networks, and computer systems to ensure the correct resources are deployed to the correct location in a timely manner. They also ensure maximum impact is achieved quickly and effectively. As a result of the increased awareness of terrorist-borne threats, such services are becoming increasingly important. An ideal terrorist attack model on a major metropolitan area would be for a terrorist group to deploy a cyber attack against key cyberinfrastructure systems (i.e. communication, and data networks), and then implement a terrorist attack against that community.

The EM community within that metropolitan area would most likely be incapable of offering an effective/coordinated response because of the crippled communication services. Since communication technology is used to coordinate personnel, resources, and analysis of situation awareness within the emergency theater (Freudenburg, 2009; Shaw, Sharma, & Takeuchi, 2009).

Post-EM brings the difficult task of restoring the affected area to its previous state through resource and personnel deployment. These tasks focus on rebuilding. The key questions the affected area, EM community must confront is who makes the decisions, how to make them, and how to restore indispensable infrastructure such as power, water, transportation, data, and communication services. Many in the EM community call this time "the window of opportunity", to build better and mitigate future risk associated with disasters or emergencies. These steps are normally the mitigative measures that would be unpopular with citizens such as better building codes, reformatting of existing infrastructure systems such as power grids and transportation systems. This is also the time to deploy important cyber security aware infrastructure systems within cyberinfrastructure systems such as intrusion detection monitors, stronger network security systems and cyber awareness campaigns to the local population (Hong & Lindu, 2009; Howitt, Leonard, & Giles, 2009; Miller, 2009).

4. Emergency management and education (cyber security)

The revolution in EM has occurred, although, most of the EM information-gatekeepers within the classroom have not yet changed their curriculum to reflect this reality. The challenge is not simply a curriculum problem, albeit it is a large and significant issue. Serious modern EM interdisciplinary issues abound, are in fact some of industries, and the US most important challenges (Plant, Arminio, & Thompson, 2011).

Many EM agencies and industries are looking for capable graduates within cyber security and EM experience, find they lack the capability to address the complex and challenging nature of interdisciplinary work, which expands beyond their traditional training in EM (Clement, 2011; Radvanovsky & McDougall, 2010). The scale and variety of the collaborative, cross-discipline interplay are not represented in traditional EM curricula (Haller, Merrell, Butkovic, & Willke, 2011).

Educational research provides strong evidence that active and collaborative learning environments provides students a much deeper and more integrated understanding of concepts as well as overall retention in courses. Successful sharing of course content, resources enable students in such courses to experience dissimilar teaching styles, which supplements diverse cognitive learning styles such as visual, auditory and kinesthetic (Caldwell, 2011).

For example, most EM programs around the country lack any virtualization tools for cyber security situational awareness. Cyber security is a notoriously challenging subject for students to comprehend. But by making extensive use of tangible artifacts such as cyber security virtualization tools to enhance the learning experience of students and teaching effectiveness of instructors, cyber security concepts could be made accessible to a more diverse student population (Bullen, Abraham, Gallagher, Simon, & Zwieg, 2009).

Furthermore, attacks on critical infrastructure can have devastating consequences these infrastructures are considered to be high-value targets for cyber terrorists. Truly modeling, and effectively demonstrating this within the context of a standard course on cyber or network security or EM training can be problematic, although, with the use of tangible tools such as simulation software, students could more easily understand the complexities of such problems. Students could interact with virtual representations of cities, counties, and nations to demonstrate cyber security attacks, and allow them to deploy solutions in real time. Thereby, enabling the students to examine existing cyber security problems within the domain of EM, which would integrate in the theoretical and practical components. Also, given contemporary students' fondness for multimedia styles of presentation, the virtualization approach would serve as a tool to combat students understanding of the unique problems related to cyber security and EM (Caldelli, Amerini, Picchioni, De Rosa, & Uccheddu, 2009; Pan & Xu, 2010; Smith & Agarwal, 2010).

As in the case above with the intrusion attacks on key critical infrastructure locations such as military installations in the United Kingdom, intrusion attacks are considered to be one of the most common types of cyber security threats facing the EM community (Jamieson, Land, Smith, Stephens, & Winchester, 2009).

5. Emergency management and intrusions

Most practitioners within the EM community wouldn't know what an intrusion is or how to handle such an incident. A recent survey of Intrusion Detection Systems (IDS) indicates that most practitioners are still examining the central question of how to best implement reactive IDS. (Allen, 2000; Arvidson & Carlbark, 2003; Escamilla, 1998; Koziol & Safari Tech Books Online., 2003; Rehman & Safari Tech Books Online., 2003; Valdes & Zamboni, 2006). In a modern cyberinfrastructure world, nodes (e.g. networked computer systems or devices) within a system maybe connected to thousands or millions of other nodes resulting in millions of possible candidates for intrusion attacks from a single or a multistage attack (Liu, Zang, & Yu, 2005). Modern reactive IDS responses to intruders include log-off an offender or modify firewall setting to block network traffic from a malicious source. Although, these approaches do not work with multistage intrusion attacks, in which an intruder will perform multiple attacks at different points. Most modern reactive IDS are based on the Denning (Denning, 1987) model in which a system monitors a system's log or audit records.

As a result, by the time that an intruder appears in a log or audit record, the event has already taken place, and for this reason, intruders particularly those committing multistage attacks take extraordinary measures to ensure their actions go unrecorded. Snort is the most commonly used IDS used within cyberinfrastructure environments (Chakrabarti, Chakraborty, & Mukhopadhyay, 2010), it can generate thousands of alerts per hour. These include sensor events, which are compared against signatures of common similar attacks, or it may build a database of temporal behavioral patterns. These approaches suffer from a high false-alert rate, which increases the overall workload of most system administrators, which has led to many administrators being weary of using reactive approaches for automatic response. The question for many administrators is when to sound the alarm to law enforcement communities, which may need to be aware of larger attacks such as cyber terrorists (Jones & Michael, 2010; Warren, 2008).

While network-based IDS cover multiple nodes with sensors. These sensors capture, and analyze the content of packets that flow through the network, although, most contemporary IDS are unable to examine encrypted packets or handle large volumes of traffic as in the case of cyberinfrastructure-oriented environments. Moreover, network-based IDSs tend to be poorly placed to detect malicious intruders who act from the inside (Ayd\ et al., 2009).

In contrast host-based IDS are not encumbered by encrypted packets since they monitor all host activities by analyzing each individual application's system calls, logs, and file modifications, while constantly monitoring the host's state. Though one significant fault in most host-based IDS engines design is their reliance on the underlying network to pass them generated events, which can become the target of the intruder, as well as degrading the overall performance of the system on which they reside. The ideal IDS would incorporate automatic protection services, which would be defined by the administrator based on intrusion types and potential impact to the system. This research will focus its efforts on using a host-based IDS Snort.

Most IDS including Snort only examine a subset of all intrusion data including connections, namely those that violate the security policy or triggered an alarm. Which, result in a very limited amount of knowledge contained within the standard log file, such as which machines are present within the network, how were they impacted by the attack. To truly understand intrusion concerns and their impact on critical infrastructure locations, the EM community would need a schema that quantifies attacks and provides a domain independent framework which makes it ideal for quantifying security threats from a universe of known security threats (Umberger & Gheorghe, 2011).

5.1 Emergency Management and Intrusion Detection

A simple approach EM practitioners could deploy, and has a proven track record is the Boyd's observe-orient decide act (OODA) model (Boyd, 1996). In 1995, a retired Air Force colonel, John Boyd an expert on military strategy, studied dogfights from the Korean, Vietnam wars respectively, and develop a strategy for advanced decision making in situations between numerous adversaries. Boyd's observe-orient decide act (OODA) model was never published in a formal sense in a book or paper, but was presented to influential politicians, civil servants and military officers, the model is currently implemented in numerous organizations, such as NATO for the monitoring and control of military operations. The model has also, been leveraged in a number of commercial companies. The model offers significant potential within the sphere of intrusion detection/EM, despite the fact it was never published in a formal scientific sense, although it has received significant extensive scientific examination through peer review analysis (Grant et al., 2007). Boyd's model was a cyclic process model of four processes interacting with the surrounding environment.

This model was based on a concept known as *tempo* i.e. that is the decision cycle time, which Boyd believed was the rhythm of the response to events. The rhythm referred to tempo of decision making, according to Boyd "in order to win, we should operate at a faster tempo or rhythm than our adversaries or, better yet, get inside the adversary's Observation-Orientation-Decision-Action loop". However, the OODA model itself does not express his concept of tempo. The four processes of the model are observe, orient, decide, and act.

Observe:

The observe is the process of acquiring information about the environment by interacting with it, sensing it, or receiving messages about it. Observation also receives internal guidance and control from the orient process, as well as feedback from the decide and act processes.

Orient:

The process of orient is the process of representing the world, based on interactive process of implicit cross-referencing, correlations interactions with unfolding circumstances. The orient process forms the way the world is observe, decide, and act i.e. situation awareness.

Decide:

The decide process is the procedure of making choices among hypotheses about the current situation and possible responses to it. Decide is guided by internal feedback from orient, and provides internal feedback to observe.

Act:

The act process is testing the chosen hypothesis by interacting with the environment. Act receives internal guidance and control from the orient process, as well as feed- forward from decide. It provides internal feedback to observe.

The EM community could modify OODA as an environment to develop automatic responses to intrusions in critical infrastructure locations. Therefore, an intrusion attack could be represented as a rational reconstruction model resulting in the OODA-RR in which each node will possess two knowledge bases:

1. One for assessing the situation (Orienting)
2. The other for deciding on the response (Deciding).

The knowledge base for the intrusion response approach could be formed using quantified weights developed by situational rules, which are extracted from national assessments, and the importance of the location i.e. its critical importance to the nation or local community.

Although, some of the challenges of EM security engineering practices include globalization of asset protection, rapid response time requirements, responsiveness to changing network infrastructure environments, and heterogeneous computing platforms. These problems are not easily solved.

A good example of a governmental agency that has taken on similar problems and developed a real tangible solution is the department of defense (DOD). The frontier of cyberinfrastructure protection is of such significance that the DOD, established the U.S. Cyber Command (USCYBERCOM), in 2009, under the US Strategic Command, the USCYBERCOM, which has the unique mission within the DOD of planning, coordinating, synchronizing, activities to direct the operations and defense of DOD cyberinfrastructure resources. As the DOD implements comprehensive cyberinfrastructure protection program, the overarching issue of detecting, protecting against unauthorized access to systems still remains the unresolved issue within all facets of DOD cyberinfrastructure resources (i.e. computer network defense (CND)) (Di Pietro, Mancini, & SpringerLink (Online service), 2008; Krutz & Vines, 2008; Mancini, Pietro, & SpringerLink (Online service), 2008; Volonino, Anzaldua, & Godwin, 2007; Zamboni, Kruegel, & SpringerLink (Online service), 2006)

6. Conclusion

In this book chapter, we discussed several cyber security concerns for the EM community. Each set of EM concern has its own unique implementation concern and characteristics. Many of the EM cyber security concerns listed in this book chapter will demonstrate a clear pattern of duplication of cyber security concerns for the entire EM community. Most EM researchers agree that there is no real killer solution to integrate in cyber situational awareness for the EM community but instead there is a real need for standards to be integrated into the EM paradigm as it currently stands. This will be evident from the cyber security concerns described in this chapter. Hence this lack of coherent knowledge offers many opportunities for further research into how to guide EM community to a framework that integrates in cyber situational awareness and develops an appreciation for cyber security concerns for each particular task within the domain of EM. Therefore, many solutions must be brought to bear on the problem.

6.1 Education

In the United States, critical infrastructure is particularly difficult to secure with standard security approaches because it is massive, distributed, and interdependent and often needs to be accessible to diverse populations. Further complicating cyber security issues in the United States, is the multiple public and private entities now collaborating to build, run and maintain this critical infrastructure. Because of these numerous threats the UShas become aware of the urgent need to educate a computing/communication security, EM capable workforce quickly, and effectively to confront these growing threats. The current cyber security/EM workforce does not reflect the unique diversity of the US, many segments of the population have been left on the sideline in this new cyber war.

The EM community of practice, which currently exists, must embrace the changing role of cyber security as a key component or task within their community. Many new cyber security and EM programs do not practice curriculum reuse or curriculum sharing. A much deeper examination of the issue demonstrates little evidence that curricular innovations are ever adopted rapidly or widely outside of their home institution or local discipline on any consistent basis. The researchers Verscoustre and McLean (Vercoustre & McLean, 2005) offers some key potential obstacles in implementing reuse, including locating the material, discovering what material is included, understanding the instructional structure and content needed to support or supplement the material and the arduous process of incorporating the content into one's course and the program curriculum. This approach must change within the homeland security disciplines.

6.2 Communal tools

Homeland security today depends as never before upon ease of access to data, associated sophisticated tools and applications, to enable asset protect, training, law enforcement.. Homeland security officers, emergency managers, police, fire departments, national security agencies who once worked in local, isolated silos now collaborate routinely and on a global scale. Specialized instruments that were spread across multiple locations can now fit into a single location connected via cyberinfrastructure resources. Set within this evolving cyberinfrastructure, networks have become the primary artery connecting homeland security individuals to each other and to the data so critical to their work. Going forward, such networks are likely to evolve to become "data mediums" where data can be positioned to serve an ever-changing tool for homeland security. The current structures of cyber security threat ensure they must be address by many facets of homeland security.

When the Internet was created, the end-to-end principle was adopted based on the assumption that the end users (mostly engineers and researchers at the time) were willing to behave cooperatively and with trust of one another. Security was not considered important to the designers. The Internet protocols and architecture were designed from the perspective of functionality. To support emerging applications, the intermediate network was a purely transparent carrier optimized for *best-effort* packet forwarding. Today, however, the Internet is operated in an untrustworthy world and any device connected to it can become a victim. As a result law enforcement individuals must have the tools to model, and predict possible threads before they happen (i.e. robust and intelligent network infrastructure). It is mandatory to detect and counteract attacks inside the core infrastructure. For example, within homeland security and EM community, it is important to make the distinction between infrastructure security and information security. Individuals steal information all of the time from agencies and industries, with types of intrusions. While when individuals target cyberinfrastructure, they are mainly targeting the availability, reliability, and stability of the network fabric.

As presented above in section 5.1, an individual could deploy a simple intrusion attack by flooding a server with data, which simply exhausts certain critical resources, such as bandwidth. The attacker does not even need to understand the fundamentals of the system. But when the attacker (s) targets large groups of systems with the goal taking down key infrastructure assets, the results can have large-scale societal implications. The EM community should adopt more nontraditional educational models to expose students within both the cyber security and EM communities to each other's disciplines. Both communities could benefit from the direct use of virtualization teaching tools such as visualization. By utilizing the proposed instructional model the traditional whiteboard classrooms could be replaced by active communal learning environments. These environments could incorporate practical/interdisciplinary computer security theories, principles, and EM tasks, which enable students to examine existing problems in innovative and unique ways, thereby, allowing them to become active participants in the learning process. As a consequence, the students' work could become a part of the learning experience of the class, and an enriching component of teaching. As well as, allow student to consider new innovative approaches to problems.

6.3 Emergency Management and The Road Ahead

There are some cases where knowledge can only be gained through trial and error. Though this method is not very efficient, it has proven itself to be one of the most effective ways to obtain useful information. However, where there is information of a sensitive nature involved, the defending actor is often reluctant to welcome would-be cyber terrorists to assault their systems. The use of the trial and error method often results in failure. This is very unappealing to many within the EM community, as failure to protect one's system can have catastrophic consequences. It then becomes necessary to create a safe environment to test one's system security against a large quantity of various types of attack. With the risks accompanying failure abolished, every iteration of the attack simulation may produce beneficial data regardless of whether or not the defenses were successful in thwarting the attack.

The defender would be well equipped, and able to react to the exploits of the actors in several different ways, with the goal of slowing and eventually stopping the attack.

Resilience:

- The first line of defense will be a sturdy firewall, and a steady stream of updates and patches. This alone will hinder actors to a limited degree.
- The resilience of any system is one of the most important aspects of system security. It acts as a preventive measure against recreational actors possessing all levels of skill and quantities of resources, and non-state organizational actors may find an especially resilient system to be a devastating deterrent.
- The patches and updates will either be automatic or applied by an administrator.

Denial:

- Denial of service can be a very effective means of deterring an actor. The repeated termination of a connection may force an attack to an abrupt end. The defender will possess the means to cause such interruptions.
- Denial is the next best thing to possessing a resilient system. Actors operating with but a few terminals may find their efforts to be in vain once they have been denied on all fronts.
- An actor with multiple terminals or networks may circumvent the denial, but the denial can also be repeated.
- This methods holds to be most effective against recreational hackers, who usually only have one viable connection to the Internet, and non-state hackers, who may have more than one connection but are still hindered by limited resources.

Retaliation:

- Retaliation should only be used as a last resort, and even then, it should be used with extreme caution. Attacking the actor may stop the attack, but the dilemma of attribution makes this method highly unreliable.
- Retaliation may serve a purpose in dealing with state actors. State actors have something to protect, and may think twice about taking aggressive action if they know that they stand to lose more than they gain.
- The use of retaliation will be readily available to the defender, but the risk of misattribution will also be present in some fashion.

Speaking candidly, it is crucial that we place greater efforts into research and development while also taking the initiative to thoroughly educate the public on the issues regarding future security of the modern society as these issues relate to EM.

7. References

Abbott, E. B., Hetzel, O. J., & American Bar Association. Section of State and Local Government Law. (2010). *Homeland security and emergency management : a legal guide for state and local governments* (2nd ed.). Chicago, IL: Section of State and Local Government Law, American Bar Association.

Allen, J., Christie, A., Fithen, W., McHugh, J., Pickel, J., and Stoner, E. (2000). *State of the Practice of Intrusion Detection Technologies.* Pittsburgh: Carnegie Mellon University.

Amin, S., & Goldstein, M. P. (2008). *Data against natural disasters : establishing effective systems for relief, recovery, and reconstruction.* Washington DC: World Bank.

Amin, S., Litrico, X., Sastry, S. S., & Bayen, A. M. (2010). *Stealthy deception attacks on water SCADA systems.* Paper presented at the Proceedings of the 13th ACM international conference on Hybrid systems: computation and control.

Arvidson, M., & Carlbark, M. (2003). *Intrusion Detection Systems -- Technologies, weaknesses and trends.* Linköping University, Stockholm.

Asimakopoulou, E., & Bessis, N. (2010). *Advanced ICTs for disaster management and threat detection : collaborative and distributed frameworks.* Hershey, PA: Information Science Reference.

Ayd\, M. A., \#305, Zaim, A. H., G\, K., \#246, & Ceylan, k. (2009). A hybrid intrusion detection system design for computer network security. *Comput. Electr. Eng., 35*(3), 517-526.

Bayer, U., Kirda, E., & Kruegel, C. (2010). *Improving the efficiency of dynamic malware analysis.* Paper presented at the Proceedings of the 2010 ACM Symposium on Applied Computing.

Becerra-Fernandez, I., Madey, G., Prietula, M., Rodriguez, D., Valerdi, R., & Wright, T. (2008). *Design and Development of a Virtual Emergency Operations Center for Disaster Management Research, Training, and Discovery.* Paper presented at the Proceedings of the Proceedings of the 41st Annual Hawaii International Conference on System Sciences.

Becerra-Fernandez, I., Xia, W., Gudi, A., & Rocha, J. (2007). *Task Characteristics, Knowledge Sharing and Integration, and Emergency Management Performance: Research Agenda and*

Challenges. Paper presented at the 16th International Conference on Management of Technology.

Bhavanishankar, R., Subramaniam, C., Kumar, M., & Dugar, D. (2009). *A context aware approach to emergency management systems*. Paper presented at the Proceedings of the 2009 International Conference on Wireless Communications and Mobile Computing: Connecting the World Wirelessly.

Boyd, J. (1996). *The Essence of Winning and Losing*: Unpublished lecture notes.

Bullen, C. V., Abraham, T., Gallagher, K., Simon, J. C., & Zwieg, P. (2009). IT workforce trends: Implications for curriculum and hiring. *Communications of the Association for Information Systems, 24*(1), 9.

Caldelli, R., Amerini, I., Picchioni, F., De Rosa, A., & Uccheddu, F. (2009). Multimedia forensic techniques for acquisition device identification and digital image authentication. *Handbook of Research on Computational Forensics, Digital Crime and Investigation: Methods and Solutions*.

Caldwell, S. L. (2011). *Critical Infrastructure Protection: Update to National Infrastructure Protection Plan Includes Increased Emphasis on Risk Management and Resilience*: DIANE Publishing.

Chakrabarti, S., Chakraborty, M., & Mukhopadhyay, I. (2010). *Study of snort-based IDS*. Paper presented at the Proceedings of the International Conference and Workshop on Emerging Trends in Technology.

Chandler, R. C., & BCP Media. (2005). *Crisis communication planning : sustaining effective corporate communication during disasters, emergencies, and critical events*. St. Louis, Mo.?: Richard L. Arnold.

Chen, R., Sharman, R., Rao, H. R., & Upadhyaya, S. J. (2008). Coordination in emergency response management. *Commun. ACM, 51*(5), 66-73.

Clement, K. E. (2011). The Essentials of Emergency Management and Homeland Security Graduate Education Programs: Design, Development, and Future. *Journal of Homeland Security and Emergency Management, 8*(2), 12.

Cohen, N. (2009). *Emergency communications : enhancing the safety network*. Hauppauge, N.Y.: Nova Science Publishers.

Collins, H. (2011). California May Incorporate Cyber-Readiness into State Emergency Plan. *Emergency Management*

Davenport, T., & Pruask, L. (1998). *Working Knowledge: How Organizations Manage What They Know*. Boston: Harvard Business Press.

Davis, L. M., Advisory Panel to Assess Domestic Response Capabilities for Terrorism Involving Weapons of Mass Destruction (U.S.), United States. Dept. of Defense. Office of the Secretary of Defense., & National Defense Research Institute (U.S.). (2006). *Combating terrorism : how prepared are state and local response organizations?* Santa Monica, CA: RAND National Defense Research Institute.

Denning, D. E. (1987). An Intrusion -Detection Model. *IEEE Transactions on Software Engineering, 23*(12), 800 - 807

Desourdis, R. I. (2009). *Achieving interoperability in critical IT and communication systems*. Boston: Artech House.

Di Pietro, R., Mancini, L. V., & SpringerLink (Online service). (2008). Intrusion detection systemspp. xiii, 249 p.). Available from http://eresources.lib.unc.edu/external_db/external_database_auth.html?A=P%7C F=N%7CID=24%7CREL=AAL%7CURL=http://libproxy.lib.unc.edu/login?url=htt p://dx.doi.org/10.1007/978-0-387-77265-3

Disaster planning and relief. Part 2. (2010). New Delhi Washington, D.C.: Library of Congress Office; Library of Congress Photoduplication Service.

Dudenhoeffer, D. D., Permann, M. R., Woolsey, S., Timpany, R., Miller, C., McDermott, A., et al. (2007). *Interdependency modeling and emergency response.* Paper presented at the Proceedings of the 2007 summer computer simulation conference.

Elmagarmid, A. K., Samuel, A., & Ouzzani, M. (2008). Community-Cyberinfrastructure-Enabled Discovery in Science and Engineering. *Computing in Science & Engineering, 10*(5), 46-53.

Escamilla, T. (1998). *Intrusion detection : network security beyond the firewall.* New York: John Wiley.

Feng, Y.-H., & Lee, C. J. (2010, 20-23 April 2010). *Exploring Development of Service-Oriented Architecture for Next Generation Emergency Management System.* Paper presented at the Advanced Information Networking and Applications Workshops (WAINA), 2010 IEEE 24th International Conference on.

Forrest, S., Hofmeyr, S. A., & Somayaji, A. (1997). Computer immunology. *Commun. ACM, 40*(10), 88-96.

Freudenburg, W. R. (2009). *Catastrophe in the making : the engineering of Katrina and the disasters of tomorrow.* Washington, DC: Island Press/Shearwater Books.

Grant, T. J., Venter, H. S., & Eloff, J. H. P. (2007). *Simulating adversarial interactions between intruders and system administrators using OODA-RR.* Paper presented at the Proceedings of the 2007 annual research conference of the South African institute of computer scientists and information technologists on IT research in developing countries.

Haller, J., Merrell, S. A., Butkovic, M. J., & Willke, B. J. (2011). Best Practices for National Cyber Security: Building a National Computer Security Incident Management Capability, Version 2.0.

Hansen, L., & Nissenbaum, H. (2009). Digital disaster, cyber security, and the Copenhagen School. *International Studies Quarterly, 53*(4), 1155-1175.

Harknett, R. J., Callaghan, J. P., & Kauffman, R. (2010). Leaving Deterrence Behind: War-Fighting and National Cybersecurity. *Journal of Homeland Security and Emergency Management, 7*(1), 22.

Hart, G., Rudman, W. B., Flynn, S. E., & Council on Foreign Relations. Independent Task Force on Homeland Security Imperatives. (2002). America still unprepared, America still in danger Available from http://www.cfr.org/publication.html?id=5099

Hong, T., & Lindu, Z. (2009, 19-21 May 2009). *Knowledge Management System of Intercity Emergency Decision Making.* Paper presented at the Software Engineering, 2009. WCSE '09. WRI World Congress on.

Howitt, A. M., Leonard, H. B., & Giles, D. (2009). *Managing crises : responses to large-scale emergencies*. Washington D.C.: CQ Press.

Jamieson, R., Land, L., Smith, S., Stephens, G., & Winchester, D. (2009). CRITICAL INFRASTRUCTURE INFOMATION SECURITY: IMPACTS OF IDENTITY AND RELATED CRIMES.

Jennex, M. E. (2008). Cyber War Defense: Systems Development with Integrated Security. *Cyber Warfare and Cyber Terrorism*, 241-253.

Jones, M., & Michael, K. (2010). Cyber terrorists'a real threat'.

Ke, T., Ming-Tian, Z., & Wen-Yong, W. (2009, 25-28 July 2009). *Insider cyber threat situational awareness framwork using dynamic Bayesian networks*. Paper presented at the Computer Science & Education, 2009. ICCSE '09. 4th International Conference on.

Kellerman, T. (2010). Cyber-Threat Proliferation: Today's Truly Pervasive Global Epidemic. *Security & Privacy, IEEE, 8*(3), 70-73.

Koziol, J., & Safari Tech Books Online. (2003). Intrusion detection with Snortpp. xx, 340 p.). Available from http://ezproxy.library.arizona.edu/login?url=http://proquest.safaribooksonline.com/?uiCode=uariz&xmlId=157870281X

Krutz, R. L., & Vines, R. D. (2008). *The CEH prep guide : the comprehensive guide to certified ethical hacking*. Indianapolis, IN: Wiley.

Lansford, T. (2010). *Fostering community resilience : homeland security and Hurricane Katrina*. Burlington, VT: Ashgate.

LearningExpress (Organization). (2010). *Becoming a homeland security professional*. New York: LearningExpress.

Liu, P., Zang, W., & Yu, M. (2005). Incentive-based modeling and inference of attacker intent, objectives, and strategies. *ACM Trans. Inf. Syst. Secur., 8*(1), 78-118.

Mancini, L. V., Pietro, R., & SpringerLink (Online service). (2008). Intrusion Detection Systems, Advances in Information Security, 38. Available from http://eresources.lib.unc.edu/external_db/external_database_auth.html?A=P%7CF=N%7CID=24%7CREL=AAL%7CURL=http://libproxy.lib.unc.edu/login?url=http://dx.doi.org/10.1007/978-0-387-77265-3

Maughan, D. (2010). The need for a national cybersecurity research and development agenda. *Commun. ACM, 53*(2), 29-31.

Mendon, D., Jefferson, T., & Harrald, J. (2007). Collaborative adhocracies and mix-and-match technologies in emergency management. *Commun. ACM, 50*(3), 44-49.

Miller, D. A. (2009). *Disaster response*. Detroit: Greenhaven Press.

Moore, M. (2010). *Bridging the gap : developing a tool to support local civilian and military disaster preparedness*. Santa Monica, CA: RAND.

Morris, J. (2009). *Disaster planning*. Detroit: Greenhaven Press.

Moteff, J. (2004). *Critical infrastructure and key assets: definition and identification*.

Murphy, T., & Jennex, M. (2006). Knowledge Management, Emergency Response, and Hurricane Katrina. *INTERNATIONAL JOURNAL OF INTELLIGENT CONTROL AND SYSTEMS, 11*(4), 199-208.

Neumann, P. G. (2010). Risks to the public. *SIGSOFT Softw. Eng. Notes, 35*(3), 24-32.

Pan, R., & Xu, C. (2010). *Research on Decision of Cyber Security Investment Based on Evolutionary Game Model.*

Plant, J. F., Arminio, T., & Thompson, P. (2011). A Matrix Approach to Homeland Security Professional Education. *Journal of Homeland Security and Emergency Management, 8*(2), 8.

Radvanovsky, R., & McDougall, A. (2010). *Critical infrastructure: homeland security and emergency preparedness*: CRC.

Rehman, R. U., & Safari Tech Books Online. (2003). Intrusion detection systems with Snort advanced IDS techniques using Snort, Apache, MySQL, PHP, and ACID, Bruce Perens' Open source series.pp. xii, 263 p.). Available from http://ezproxy.library.arizona.edu/login?url=http://proquest.safaribooksonline.com/?uiCode=uariz&xmlId=0131407333

Rocha, J., Becerra-Fernandez, Xia, W., & Gudi, A. P. (2009, Aug 2009). *Dealing with Task Uncertainty in Disaster Management: The Role of Knowledge Sharing for Exploration and Exploitation.* Paper presented at the Amerias Conference on Information Systems, San Francisco, California

Shaw, R., Sharma, A., & Takeuchi, Y. (2009). *Indigenous knowledge and disaster risk reduction : from practice to policy.* New York: Nova Science Publishers.

Smith, C. L., & Agarwal, R. (2010). Practicing Safe Computing: A Multimedia Empirical Examination of Home Computer User Security Behavioral Intentions. *Management Information Systems Quarterly, 34*(3), 613-643.

Umberger, H., & Gheorghe, A. (2011). Cyber Security: Threat Identification, Risk and Vulnerability Assessment. *Energy Security,* 247-269.

United States. Federal Emergency Management Agency., United States. Federal Emergency Management Agency. Community Preparedness Division., & Citizen Corps (USA Freedom Corps). (2009). *Personal preparedness in America : findings from the Citizen Corps national survey.* Washingotn, D.C.: Community Preparedness Division, FEMA.

Valdes, A., & Zamboni, D. (2006). *Recent advances in intrusion detection : 8th international symposium, RAID 2005, Seattle, WA, USA, September 7-9, 2005 : revised papers.* Berlin ; New York: Springer.

Vercoustre, A.-m., & McLean, A. (2005). Reusing Educational Material for Teaching and Learning: Current Approaches and Directions. *International Journal on E-learning (IJEL), a special issue on Technologies for Electronic Documents, 4*(1), 57-68.

Volonino, L., Anzaldua, R., & Godwin, J. (2007). *Computer forensics : principles and practices.* Upper Saddle River, N.J.: Pearson/Prentice Hall.

Warren, M. (2008). Hackers and cyber terrorists. *Encyclopedia of information ethics and security,* 304.

Watts, D. (2003). *Security & vulnerability in electric power systems.*

Wenger, E., McDermott, R., & Snyder, W. M. (2002). *Cultivating Communities of Practice: A Guide to Managing Knowledge.* Cambridge, MA: Harvard Business School Press.

Zamboni, D., Kruegel, C., & SpringerLink (Online service). (2006). Recent advances in intrusion detection 9th international symposium, RAID 2006, Hamburg, Germany, September 20-22, 2006 : proceedings, Lecture notes in computer science,pp. xii, 330 p.). Available from http://www.springerlink.com/openurl.asp?genre=issue&issn=0302-9743&volume=4219

2

Lessons Unlearnt: The (Human) Nature of Disaster Management

David Hutton
Deputy Director
United Nations Relief and Works Agency

1. Introduction

This chapter examines the impact of human behavior in disaster management and emergency preparedness. It now is recognized that disasters are rarely 'natural', but rather a product of the interface between hazards and human activity. Population growth, urbanization, poverty, and poor urban planning are human causative factors which have received considerable attention. In response, emergency managers have engaged in widespread mitigation and preparedness efforts, including both investment in technologies and systems as well as public education and awareness-raising. The results have been mixed, even in the most developed countries with advanced emergency management systems; the European heat wave in 2003 claimed over 35,000 lives while Hurricane Katrina in 2005 took over 1,800 lives in the United States.

The failure to effectively respond to disaster events is generally attributed to the shortcomings of emergency management systems, inadequate planning, poor communication and/or coordination. While these reasons are certainly valid, what is frequently overlooked is the possible role that human nature may have in perpetuating these crisis. In *Flirting with Disaster*, Marc Gerstein (2008) points out that accidents and disasters are rarely accidental. Citing disasters ranging from the Challenger and Columbia space shuttle explosions to Chernobyl and Katrina, the author points that organizations that rely on a relatively small number of experts (as is the case with many emergency response teams) can become prone to *group think* which fosters its own form of human bias, distortion, and errors of judgment. As Gerstein points out in the case of the Columbia, "[it] is the story of how organizational pressures, public relations concerns, and wishful thinking contributed to a phenomenon known as *bystander behavior* - the tendency of people to stand on the sidelines and watch while things go from bad to worse".

Emergency management in North America has its roots in civil defense and by necessity retains a strong *command and control* dimension. It has also been largely driven by associated professions including the military, police, fire and other emergency services. As such, emergency management retains a hierarchical culture grounded in clearly defined roles and responsibilities, highly functional and technical systems, and standard operating procedures. While such are critical to the effective management of crises and disaster events, research has shown that these must also be balanced with flexible decision-making, stakeholder awareness, and basic human qualities such as trust and collaboration. Without such, "there can be reinforced silos of hierarchy and structure that not only hinder communication intra- and inter-organizationally, but constrict interdisciplinary thinking, sharing and trans-departmental thinking" (Devitt & Bordzicz, 2008). With this, there may also be a devaluing of the importance of engaging and planning *with* (rather than *for*) external stakeholders including the public and communities. This can have the further affect of limiting emergency managers' understanding of those critical factors that not only make communities vulnerable but can also be built upon to enhance preparedness and resiliency.

Devitt and Borodzicz (2008) have similarly raised the importance of better understanding incident command systems and leadership styles. The authors note that current models of crisis leadership frequently fail to establish a balance between the requirement for task skills, interpersonal skills, stakeholder awareness and personal qualities. This can have significant consequences during an emergency. "Leaders managing crises under stressful situations are likely to revert to the style which they are most comfortable - an unconscious preference perhaps - and the more disturbing the situation, the stronger the urge to take refuge in familiar procedures" (Devitt & Borodzicz, 2008: 212). To this point, Legadic (1993, cited in devitt & Borodzicz, 2008) has observed, "leaders who are more task-oriented than human-relations oriented [may] reach the point where they neglect human relations altogether (and vice versa)" (212). Additionally, should subordinates be constrained by fears of breaching organizational taboos - be these cross-functional, technical, or hierarchical - it is likely that their effectiveness and usefulness to an organization will also be compromised (Robert & Lajcha, 2002).

The influence of basic human qualities such as trust can not be understated. Dekker et al. (2008), in studying the effectiveness of first responder agencies to learn from response failures, found that the least effective agencies fell short in terms of the most basic human qualities, these being mutual trust and participation. Milstein (in Ross, 2005: 3) has observed that "the development of trust is quite often the single most important tool in overcoming barriers and obstacles ... Effective communication, goal attainment, and service attainment are possible only in atmosphere of trust". Going a step further, Peterson and Besserman (2010) have emphasized the importance of trust in building and maintaining informal networks that serve to crisscross the borders of functions, hierarchies, and business units that characterize most governments and organizations. "Efficiency in response is increased since someone who is known informally and in a positive light has a greater propensity of saying yes when asked for assistance and/or resources. This leads not only to more effective response, but more efficient response as well" (Peterson & Besserman, 2010: 9).

Examples of failed emergency responses as a result of poor trust and limited communication and information sharing litter the literature of disaster and emergency management. In studying inter-governmental responses to disasters, Comfort (2002) observed that hierarchical organizations that fail to account for such factors often breakdown due to a lack of timely information flow and analysis, constraints on innovation, an inability to rapidly shift resources, as well as difficulty in responding to new and/or unexpected demands. In the case of the Challenger explosion, for example, Gerstein (2008) points out that a combination of organizational pressures, public relations concerns, and wishful thinking led managers to overlook the o-ring risks voiced by subordinates and launch of the shuttle. In the case of September 11th, investigations revealed a serious problem in the sharing of information between government organizations, which in turn compromised the capacity of the government to detect and respond to the terrorist attack (9/11 Commission, 2004). In reference to FEMA's over cautiousness and delayed response to the humanitarian crisis of Hurricane Katrina, Sobel and Leeson (2006: 59) have argued that this in part can be traced to a "reluctance to trust local officials due to the widely-held perception of rampant public-sector corruption in New Orleans (and the State of Louisiana)".

In part, these incidents reflect the inherent risks associated with the emergence of 'virtual teams' and organizational processes that increasingly rely on technologies and software to ensure coordinated planning, information management, decision-making and communications. While these functions are critical to managing complex operations, over-reliance on 'systems' may contribute to a second set of problems associated with human nature. In many cases, agencies may have specific mandates that contribute to 'stovepipe' operations, or rigid functionally organized departments that act as 'silo traps' for information (Eggers & O'Leary, 2009). When interactions do occur, especially when not face-to-face, "cultural and language differences become magnified, as do conflicts. It is much easier to hide errors and problems, sweep misunderstandings under the rug, and make erroneous assumptions when you are communicating via phone and e-mail rather than person. Furthermore, such mistakes and mix-ups are more likely to become full-fledged disasters when the group does not feel free to acknowledge and address them openly" (Ross, 2005: 3).

To this point, Delorme (personal communiqué, September 5, 2011) has observed that many incident command courses and trainings do not focus on leadership competencies or the development of these competencies with participants. Rather, the focus is typically on the procedures and processes of emergency response systems. This leads to "poor" leaders reverting and requiring adherence to the processes of incident management systems rather than the prima fascia issues of the emergency (namely information sharing, coordination, and tem work). This factor becomes more problematic as the emergency (and therefore the emergency response system itself) increases in complexity, requiring greater coordination, information sharing, and joint decision-making across agencies. In such instances, the importance of incident command systems and processes often decrease vis-à-vis the importance of leadership traits of an incident commander. These latter typically include a willingness to receive and share information, to rely on sources and expertise outside of one's own organization, to work in coordination with other organizations, and to view the achievement of strategic and tactical objectives as an integrated team effort (Comfort, 2002; Currao, 2009).

Since the 1990s, especially within the corporate sector, the importance of 'soft' leadership skills (or emotional intelligence) has been recognized as being critical to both effective day-to-day management but also crisis management. Emotional intelligence can be broadly defined as an ability to recognize the meanings of emotion and their relationships, and to reason and problem-solve on the basis of these (Mayer et al., 1999). Elements of emotional intelligence include self-awareness of one's strengths and weaknesses (and how these affect others), self-regulation of disruptive emotions and impulses, motivation to achieve beyond expectations, empathy or understanding or other's feelings, and social skills to engage others and manage relationships (Goleman, 1998).

Goleman (1998) has argued that while intelligence, toughness, determination and vision are required for success, these are not sufficient in and of themselves to achieve the highest levels of achievement. Further, emotional intelligence becomes increasingly relevant and important in senior management positions which require more leadership than technical skills. "It is not that IQ and technical skills are irrelevant. They do matter, but mainly as *threshold capabilities*; that is, they are the entry-level requirements for executive positions ... [But] emotional intelligence is the sine qua non of leadership. Without it, the person can have the best training in the world, an incisive, analytical mind, and an endless supply of smart ideas, but still won't make a great leader" (Goleman, 1998: 3).

Applied to emergency management, the concept of emotional intelligence as a premise for building trust, communication has obvious implications. As noted in previous examples, including September 11th and Hurricane Katrina, the breakdown in both preparedness and response activities can often reflect a clear lack of trust between agencies and different levels of government. However, the concept of emotional intelligence can be applied at a more basic level, where nuances of human behavior may be more subtly manifested in styles of communication and decision-making styles that may have immediate and tragic results. As one such example, the main factor leading to the 1978 United Airlines DC-8 crash in Portland, Oregon was attributed to the failure of the crew members to successfully communicate concerns to the captain that the plane was running low on fuel (National Aviation Safety Board, 1978). In *Outliers: The Story of Success*, Malcolm Gladwell cites the crash of Korean Air Flight 801 as an example of how cultural differences can lead to accidents. In the case of the Korean flight, it continued to circle the Guam airport at the request of the controller while running out of fuel, a decision which Gladwell attributes to a culture's Power Distance Index (P.D.I.), this being a measurement of "how much a particular culture values and respects authority".

At the same time, at a much broader level, there is a need to apply the concepts of trust, collaboration and emotional intelligence to the strategies by which emergency managers engage the public and the communities they serve. In part because emergency management has its origins in the Cold War, filled largely by emergency services professionals with strong command and control backgrounds, the public has frequently been perceived as a *problem to be solved* rather than *part of a solution* to disasters. Schooch-Spana (2004), for example, has observed that exercises testing emergency response capabilities to biological attacks have frequently framed the public as mass casualties or hysteria-driven mobs. Additionally, "public communication and risk communications have become code words with which to skirt the multifaceted realities associated with community response to terrorism, bio-terrorism in particular. When authorities say they want better communication with the public, what they [often] tend to mean is that they want public buy-in, compliance, and understanding - possibly even absolute - when tough choices arise (e.g., how to distribute scarce resources in an emergency" (Schooch-Spana, 2004: 2).

This attitude of a 'problematic' public is not only potentially pejorative but does not adequately reflect the complexity and underlying reasons as to why people do not adequately prepare for emergencies. Terms such as *apathy*, *denial* and *avoidance* are found throughout emergency preparedness literature to describe why households continue to be unprepared for disasters. Marsha Evans, an American Red Cross president, has used five amusing but not necessarily enlightening terms to describe the 'unprepared' public: "head scratchers" who don't know where to find preparedness advice; "head in the sand" types who believe preparation is unimportant; "head in the clouds" people who mistakenly believe they are ready; the "headset crowd" that is too busy and can't find time to do it; and people who "simply haven't thought about preparedness" (Mintz, 2004).

These explanations, while perhaps applicable to some people, fail to capture both the nuances and complexity of human behavior. Indeed, the very concept of denial or normalcy bias (a tendency for people to underestimate the possibility of a disaster occurring and its possible effects) is also an essential adaptive mechanism by which people are able to cope with the myriad of stressors they face in life, including serious illness, the lost of loved ones and other tragedies they may endure.

In part, this reflects the tendency of emergency managers to plan for disasters within their own silos, outside the context of other life demands. Framed within the context of emotional intelligence, insufficient attention and understanding (in other words, *empathy*) has been given to the fact that many people (such as the poor or chronically ill) may have more pressing realities of daily living that make the risk of a disaster (which may or may not occur) pale in comparison. To this point, Hutton et al. (2007), in studying preparedness to risks associated with extreme weather events in Canada, found that while 80% of respondents believed the weather was changing, only 4% cited it as a personal worry, in large part because they were more concerned with more urgent daily needs such as personal or family health (32%) or financial issues (28%).

Indeed, the manner in which preparedness information is presented may often have little relevance to a targeted population. In a revealing survey of public perception and response to heat warnings across urban centres in North America, Sheridan (2006) found that while knowledge of the event was widespread (upwards of 90%), only 46% of respondents had changed their behaviour. A majority (60%) did not believe the message was meant for them while those that had changed their behaviour had done so because "it was hot", not because of the heat warning. This can to an extent be attributed to a failure among emergency managers to *understand* and *motivate* the public. To this point, Veil et al. (2009) found that emergency managers in the United States generally relied on one-way media and publicity in their effort to increase citizen emergency preparedness, conceptualizing communication primarily as the dissemination of preparedness messages rather than as a process of research and evaluation (Veil et al., 2009). This intuitively suggests that more engaging and alternative approaches to public awareness might result in higher levels of preparedness. Indeed, the American Red Cross (2006) found that much of the public is in fact not resistant preparedness; while 60% of surveyed Americans indicated they were unprepared for a disaster of any kind, 82% agreed that "If someone could make it easy for me to be prepared, I'd do it".

Despite these findings, emergency managers have been slow to adapt to new practices which might reach targeted populations in more effective ways. In part, this can be attributed to a failure to *think outside the box*, even after disasters occur and lessons may be readily drawn upon. Instead, as James (2004) has observed, organizations too often adopt a reactive and defensive position that prevents learning, focusing on damage control rather than identifying and implementing organizational change efforts aimed at reforming or strengthening organizational systems, policies or procedures. This has been similarly observed by Roux-Dufort (2007). In examining crisis management research of large scale industrial accidents (including Three Mile Island, Bhopal, Chernobyl, the Challenge and

Exxon Valdez incidents), Roux-Dufort found that that such were often descriptive and generated 'knowledge about accidents than organizations', rather than focusing on underlying organizational processes and decision-making structures that that may also serve as contributing factors. To this end, James (2004: 7) has written, "What is needed is not simply management of the situation but acts of leadership whereby the organization, crisis, and the environment are considered holistically ... Crisis leadership first involves a *corporate mindset* that allows for the possibility that forms are vulnerable to uncontrollable events *and* that there may be bad seeds in the organization that intentionally or unintentionally engage in behaviors that lead to crisis".

The consequence of not fully understanding and engaging the public can be significant. In reviewing the failures of the Katrina response in regard to supporting people with disabilities, for example, it was found that many emergency managers were simply misinformed. Many shelters did not have ramps for wheelchairs, accessible toilets for persons with disabilities, as well as alternative information formats for the visually and hearing impaired (National Council on Disability, 2006). A survey of emergency managers further revealed that only 27% of surveyed emergency managers had taken the FEMA planning course for persons with special needs, 58% did not have preparedness materials for seniors or people with disabilities, while 57% did not know the proportion of people with disabilities who were residing within their jurisdictions (Fox et al, 2005).

The solution, however, is not as straightforward as enhancing the knowledge and skills of emergency managers. Cahill (in Heller, 2007: 1) has observed that "one of the most amazing things that's been found is just how little personal disaster planning there is among people with disabilities. And it's not because the materials aren't there but too few people with disabilities use it". As an example, one year after Hurricane Katrina and Rita, only 3,000 people in Houston (an area also prone to severe storms) living with disabilities signed up with a special needs registry for services targeting people with disabilities during emergencies. After two years, this figure dropped to only 500 (Heller, 2007).

The failure of individuals to prepare for emergencies, unfortunately, is often viewed by emergency managers as acts of denial or irresponsibility. This is neither an accurate nor helpful explanation. Rather, the focus of emergency managers should be on how best to engage and *motivate* individuals and communities in activities which will enhance their capacity to adjust to unpredictable but potentially disastrous events. In fact, it is now acknowledged that simple awareness or even understanding of a possible risk is not a sufficient condition for behavioural change. Ronan and Johnston (2005: 7) have observed that *motivation*, as opposed to information and education, may be the *sine qua non* of community preparedness. "Despite the fact that people may be aware of both risk as well as strategies that can mitigate that risk, it does not follow directly that they will take the necessary action ... Motivation is the psychological factor that fuels interest, concern, and action".

A significant challenge for emergency managers, then, is how best to *motivate* and *engage* individuals and communities in activities which will enhance their capacity to adjust to unpredictable but potentially disastrous events. To this end, Conklin (2008) has emphasized the importance of involving people in finding solutions, rather than simply telling them what to do. "To put it more starkly, without being included in the thinking and decision-making processes, members of the social network may seek to undermine or even sabotage the projects if their needs are not considered" (3). Moreover, when information is not forthcoming, or from sources that are not fully trusted, invites negative stakeholder reaction. "Failure to adequately and in a timely fashion address a crisis situation gives stakeholders the opportunity to 'fill in the blanks'. In the absence of information, or the presence of poor or inadequate information, people tend to assume the worst and then base their subsequent behaviour on those negative assumptions" (James, 2004: 4). Perhaps among the more illustrative examples of this is the 1979 Three Mile Island nuclear accident, when 140,000 people evacuated the area within days as a result of conflicting and confusing messaging and communication from public officials. "What made these significant was a series of misunderstandings caused, in part, by problems of communication within various state and federal agencies. Garbled communications reported by the media generated a debate over evacuation. Whether or not there were evacuation plans soon became academic. What happened on Friday was not a planned evacuation but a weekend exodus based not on what was actually happening at Three Mile Island but on what government officials and the media imagined might happen. On Friday confused communications created the politics of fear". (Cantelon & Williams, 1982: 50)

As such, emergency managers must recognize disasters as social constructs and see people as *part of the solution* rather than *part of problem* to be solved or managed during an emergency. This requires 'people-focused' (rather than technical) planning methodologies that move beyond *planning for* to *planning with* all segments of society, including the most vulnerable and marginalized groups that are more readily overlooked. This not only begins to ensure that emergency planning and response capacities can more effectively address the diverse needs of people, but can be an important step to *engaging* and *empowering* people to better prepare themselves for emergencies and other critical events.

To achieve this, there must ultimately be a willingness to study the underlying causes of human behavior. Devitt and Borodzicz (2008: 212) have argued that "crisis leaders need to be able to put themselves in the position of all stakeholders, including the victims, and need to be able to recognize their diverse needs and feelings". As such, emergency managers must move away from a *command and control* philosophy and recognize the criticality of human behavior in shaping how both people and organizations respond to crises and emergencies. From a lessons learned perspective, this will require that managers challenge themselves to examine how their own attitudes and beliefs impact on planning and response. At the core of this, as it is at the core of enhancing organizational response capacity, is the ability to perceive the needs of others, engage in meaningful social dialogue, and motivate people to undertake activities and changes that they might otherwise not. These are essentially the hallmarks of leadership and emotional intelligence.

This is also at the core of moving towards a more coherent approach to identifying and implementing lessons that promote real change and adaptation of organizations and emergency management practices. As observed by Gerstein (2008), overtly defensive organizations and governments may be prone to *anti-learning* as a way to avoid blame and findings of faulty decision-making. As such, leaders and senior managers may also draw upon lessons learned from elsewhere, such as the Toyota *kaizen* philosophy which encourages and expects managers to continually identify problems (without fear or threat) as tool for continuous improvement (Shook, 2010). This also speaks to a basic choice in human behavior, that of a "[willingness and] ability to focus on solving problems without pointing fingers and looking to place the blame on someone" (68). This, perhaps above all else, is the most essential lesson to be learned should the field begin to fully address the human elements that contribute to lessons being unlearnt.

2. References

9/11 Commission (2004). The 9/11 Commission report: Final report of the National Commission on terrorist attacks in the United States. Accessed at: http://govinfo.library.unit.edu/

American Red Cross (2006). American Red Cross Preparedness Survey. Accessed at http://www.redcross.org/.

Cantelon, P. & Williams, C. (1982). Crisis Contained, The Department of Energy at Three Mile Island. Carbondale, Ill: Southern Illinois University Press,

Comfort, L. (2002). Managing intergovernmental responses to terrorism and other extreme events. Publius, 32 (4), 29-29.

Conklin, J. (2008). Wicked problems and social complexity. Cognexus Institute. Accessed at: http://cognexus.org/wpf/wickedproblems.pdf

Currao, T. (2009). The new role for emergency management: Fostering trust to enhance collaboration in complex adaptive emergency response systems. Master's Theses, Naval Postgraduate School, Monterey, California. Accessed at: http://www.dtic.mil/dtic/tr/fulltext/u2/a514087.pdf

Dekker, S., Jonsen, M., Bergstrom, J. & Dahlstrom, N. (2008). Learning from failures in emergency response: Two empirical studies. Journal of Emergency Management, 6 (5), 64-70.

Deverell, E. & Olsson, E. (2009). Learning from crisis: A framework of management, learning, and implementation in response to crisis. Journal of Homeland Security and Emergency Management, 6 (1), 1-20. Accessed at http//www.bepress.com/jhhsem/vol16/iss1/85

Devitt, K. & Borodzicz, E. (2008). Interwoven leadership: The missing link in multi-agency major incident response. Journal of Contingencies and Crisis Management, 16 (4), 208-216.

Disability for Empowerment, Advocacy, and Support (2006). Katrina disability information. Accessed at: http://www.katrinadisability.info/

Egger, W. & O'Leary, J. (2009). The silo trap: The wall between us. Harvard Business Press. Boson, Massachusetts.

Fox, M., White, G., Rooney, C. & Rowland, J. (2005). Nobody left behind. Research and Training center for Independent Living at the University of Kansas. Accessed at: http://www.nobodyleftbehind2.org/findings/pdfs/FProgressReport2.pdf

Gerstein, M. 92008). Flirting with disaster: Why accidents are rarely accidental. New York: Union Square Press.

Gladwell, M. (2008). Outliers: The Story of Success. London: Little, Brown & Company.

Goleman, D. (1998). What makes a leader? Best of Harvard Business Review. Harvard Business Review. Accessed at: www.hbr.org.

Heller, P. (2007). Experts urge focus on special needs. Disasters News Network, June 11. Accessed at: http://www.disasternews.net/news/article.php?articleid=3216

Hutton, D. Haque, E., Chowdhury, P. & Smith, G. (2007). Impact of Climate Change and Extreme Events on the Psychosocial Well-Being of Individuals and the Community, and Consequent Vulnerability: Mitigation and Adaptation by Strengthening Community and Health Risk Management Capacity. Report prepared for the Climate Change Impacts & Adaptation Program (CCIAP), Earth Science Division, Natural Resources Canada, Ottawa, Canada.

James, E. (2004). Crisis leadership. Technical note. University of Virginia Darden School Foundation.

Lagedic, P. (1993). Preventing chaos in a crisis: Strategies for prevention, control and damage limitation. London: McGraw-Hill.

Mayer, J., Caruso, D. & Salovey, P. (1999): Emotional intelligence meets traditional standards for intelligence. *Intelligence* 27: 267–298.

Mintz, J. (2004). Are you ready for a terrorist attack? Americans are simply not preparing for terror attack, studies find. The Washington Post, July 29. Accessed at http://healthandenergy.com/preparing_for_terrorist_attacks.htm

National Aviation Safety Board (1978). Aircraft accident report: United Airlines, Inc. McDonnell-Douglas Dc-8-61, N80820, Portland Oregon. December, 28, 1978. Report Number - MTSB-AAR-79-7. Washington, D.C: United States Government.

National Council on Disability (2006). The impact of Hurricane Katrina and Rita on people with disabilities: A look back and remaining challenges. Washington, DC: national Council on Disability.

Peterson, D. & Besserman, R. (2010). Analysis of informal networking in emergency management. Journal of Homeland Security and Emergency Management, 7 (1), 1-14. Accessed at: http://www.hepress.com/jhsem/vol7/iss1/62.

Robert, B. & Lajtha, C. (2002). A new approach to crisis management, Journal of Cotingencies and Crisis Management, 10, 181-191.

Ronan, K., & Johnston, D. (2005). Promoting Community Resilience in Disasters: The Role for Schools, Youth, and Families. New York: Springer.

Ross, J. (2006). Trust makes the team go'round. Harvard Business Review, June. Accessed at: http//hmu.harvardbusinessonline.org

Roux-Dufort, C (2007). Is crisis management (only) a management of exceptions? Journal of Contingencies and Crisis Management, 15 (2), 105-114.

Schooch-Spana, M. (2004). Biodefense: If risk communication is the answer, what is the question? Natural Hazards Observer, September, 24 (1), 1-5.

Sheridan, S. (2006). A survey of public perception and response to heat warnings across four North American cities: An evaluation of municipal effectiveness. International Journal of Biometeorology, 52 (1), 3-15.

Shook. J. (2010). How to change a culture: Lessons from NUMMI. MIT Sloan Management Review, 51 (2), 63-68.

Sobel R. & Leeeson, P. (2006). Government's response to Hurricane Katrina: A public choice
    analysis. Public Choice, 127, 55-73.

Veil, S., Littlefield, R. & Rowan, K (2009). Dissemination as success: Local emergency
    management communications practices. Publics Relations Review, 35 (4), 449-451.

3

Developing Real-Time Emergency Management Applications: Methodology for a Novel Programming Model Approach

Gabriele Mencagli and Marco Vanneschi

Department of Computer Science, University of Pisa, L. Bruno Pontecorvo, Pisa
Italy

1. Introduction

The last years have been characterized by the arising of highly distributed computing platforms composed of a heterogeneity of computing and communication resources including centralized high-performance computing architectures (e.g. clusters or large shared-memory machines), as well as multi-/many-core components also integrated into mobile nodes and network facilities. The emerging of computational paradigms such as *Grid and Cloud Computing*, provides potential solutions to integrate such platforms with data systems, natural phenomena simulations, knowledge discovery and decision support systems responding to a dynamic demand of remote computing and communication resources and services.

In this context time-critical applications, notably emergency management systems, are composed of complex sets of application components specialized for executing specific computations, which are able to cooperate in such a way as to perform a global goal in a distributed manner. Since the last years the scientific community has been involved in facing with the programming issues of distributed systems, aimed at the definition of applications featuring an increasing complexity in the number of distributed components, in the spatial distribution and cooperation between interested parties and in their degree of heterogeneity.

Over the last decade the research trend in distributed computing has been focused on a crucial objective. The wide-ranging composition of distributed platforms in terms of different classes of computing nodes and network technologies, the strong diffusion of applications that require real-time elaborations and online compute-intensive processing as in the case of emergency management systems, lead to a pronounced tendency of systems towards properties like self-managing, self-organization, self-controlling and strictly speaking *adaptivity*.

Adaptivity implies the development, deployment, execution and management of applications that, in general, are dynamic in nature. Dynamicity concerns the number and the specific identification of cooperating components, the deployment and composition of the most suitable versions of software components on processing and networking resources and services, i.e., both the quantity and the quality of the application components to achieve the needed Quality of Service (QoS). In time-critical applications the QoS specification can dynamically vary during the execution, according to the user intentions and the information produced by sensors and services, as well as according to the monitored state and performance of networks and nodes.

The general reference point for this kind of systems is the Grid paradigm which, by definition, aims to enable the access, selection and aggregation of a variety of distributed and heterogeneous resources and services. However, though notable advancements have been achieved in recent years, current Grid technology is not yet able to supply the needed software tools with the features of high adaptivity, ubiquity, proactivity, self-organization, scalability and performance, interoperability, as well as fault tolerance and security, of the emerging applications.

For this reason in this chapter we will study a methodology for designing high-performance computations able to exploit the heterogeneity and dynamicity of distributed environments by expressing adaptivity and QoS-awareness directly at the application level. An effective approach needs to address issues like *QoS predictability* of different application configurations as well as the predictability of reconfiguration costs. Moreover adaptation strategies need to be developed assuring properties like the stability degree of a reconfiguration decision and the execution *optimality* (i.e. select reconfigurations accounting proper trade-offs among different QoS objectives). In this chapter we will present the basic points of a novel approach that lays the foundations for future programming model environments for time-critical applications such as emergency management systems.

The organization of this chapter is the following. In Section 2 we will compare the existing research works for developing adaptive systems in critical environments, highlighting their drawbacks and inefficiencies. In Section 3, in order to clarify the application scenarios that we are considering, we will present an emergency management system in which the run-time selection of proper application configuration parameters is of great importance for meeting the desired QoS constraints. In Section 4 we will describe the basic points of our approach in terms of how compute-intensive operations can be programmed, how they can be dynamically modified and how adaptation strategies can be expressed. In Section 5 our approach will be contextualize to the definition of an adaptive parallel module, which is a building block for composing complex and distributed adaptive computations. Finally in Section 6 we will describe a set of experimental results that show the viability of our approach and in Section 7 we will give the concluding remarks of this chapter.

2. Related works

When a distributed application is executed on a dynamic execution environment, where the number and features of nodes and network facilities but also the desired QoS level may vary significantly and in unpredictable ways, we need to design systems which are able to modify their actual *configuration* for maintaining and respecting the desired objectives. For application configuration we intend a specific identification of application components, their mapping onto the available computing resources, the specification of their internal behaviors (e.g. if they exploit a sequential or a parallel elaboration) and the way in which the computations are performed (i.e. for a sequential component the selected algorithm whereas for a parallel computation the exploited parallelism scheme). The actual system configuration can be modified by exploiting dynamic reconfiguration activities classified in four general categories (Arshad et al., 2007; Gomes et al., 2007; Hillman & Warren, 2004; Tsai et al., 2007; Vanneschi & Veraldi, 2007):

- **Geometrical Changes**: affect the mapping between the internal structure of an application component and the system resources on which it is currently executed. Such class of changes consist in migrating a component in response to specific conditions, e.g. when a system resource fails or a new node is included in the system. These reconfigurations can be useful for load-balancing or for improving the service reliability;
- **Structural Changes**: affect the internal structure of an application component. A notable case is that of a distributed component featuring a parallel implementation in terms of multiple concurrent processes. In this situation structural changes consist in modifying the number of executed processes. This occurs when an insufficient number of nodes are available or if the task scheduler decides to change the set of processors allocated to a specific component;
- **Implementation Changes**: are modifications of the behavior of an application component, which changes completely its internal algorithm but preserving the elaboration semantics and the input and output interfaces with the other application components;
- **Interface Changes**: are intensive modifications of the component behavior, which completely changes the set of its provided external operations and services.

The existing programming models face the reconfiguration support development in two different ways. The first solution consists in putting everything concerning the adaptive behavior of a distributed application, like reconfiguration implementations, inside the run-time support system in such a way as to completely hide these aspects from the programmer standpoint. This *transparent-approach* requires a deep knowledge of the application structures so that the reconfiguration code can be automatically extracted by a static process of compilation of the program. In this scenario the programmer is freed from directly programming the reconfiguration phase but it is only involved in defining, by means of proper directives or programming constructs, the reconfigurations that are admissible (e.g. the compatibility between different versions of a component). On the other hand other programming models adopt a completely different approach to adaptivity, in which the programmer is directly involved in defining the reconfiguration activities. For each possible reconfiguration the programmer must provide the implementation that performs the dynamic reconfiguration, and it is also responsible for performing these activities in a fully correct and consistent manner.

Based on the guidelines provided above, in the rest of this section we will introduce some interesting programming models and frameworks for distributed environments. In the first part we will introduce frameworks for developing adaptive distributed applications for mobile and pervasive systems, whereas in the second part we will concentrate on programming models and tools for adaptive high-performance computations.

2.1 Exploiting adaptive behaviors in mobile and pervasive environments

Some research works provide programming frameworks for mobile applications that adapt their behavior in response to the actual level of resource availability (especially of communication networks). For instance in **Odyssey** (Noble et al., 1997) mobile applications exploit run-time reconfigurations which are noticed by the final users as a change in the application execution quality. The Odyssey operating system is responsible for exploiting periodically resource monitoring activities and for interacting with mobile applications raising or lowering their corresponding quality level. In this approach all reconfiguration actions are automatically triggered by the run-time system without any direct user intervention. Nevertheless the most important drawback of this approach is the quite limited definition of the quality concept: in many cases it only consists in the quality of the visualized data but this assumption can be restrictive when we consider more complex applications involving an intensive cooperation between computation, communication and visualization phases.

In past researches, adaptivity consists in migrating specific parts of a distributed computation for optimizing performance and energy parameters as well as for reliability reasons. As an example in **Aura** (Garlan et al., 2002) adaptivity is expressed introducing the abstract concept of *task*: i.e. a specific work that a user has submitted to the system which can be completed by several distinct applications suitable for different classes of computing resources (e.g. workstations and smartphones). The framework is responsible for exploiting migration activities in a fully transparent way w.r.t the final users. Unfortunately this approach has been developed for very simple user tasks (e.g. writing a document or preparing a presentation). On the other hand, if we consider more complex mobile applications (e.g. that involve compute-intensive computations), transferring a partially computed task to a different platform can be a critical issue concerning both performance and consistency issues that have not been analyzed in these frameworks.

Recently the execution of high-performance parallel applications on mobile computing platforms have been addressed in some preliminary research works. A relevant example is **MB++** (Lillethun et al., 2007), a framework for developing compute-intensive programs for mobile and pervasive environments. In this approach one of the most important shortfalls is the quite limited utilization of mobile nodes, which are limited to pre-processing or post-processing activities whereas compute-intensive elaborations are executed only on HPC resources. In many time-critical scenarios, such as emergency management systems, we often require the possibility to execute forecasting models and decision support systems on a distributed set of localized mobile resources, equipped with a sufficient computational power (e.g. multicore smartphones).

2.2 Existing programming models for high-performance adaptive applications

Parallel computations are a particular class of tightly coupled distributed applications composed of several cooperating parallel modules. Adaptivity for parallel applications executed on traditional HPC architectures is an important feature, especially for real-time processing in which the presence of strong real-time deadlines and performance constraints require the capability of automatically adapting the application configuration for quickly responding to platform changes. Emergency management systems (e.g. disaster prediction and management, risk mitigation of floods and earthquakes) are notable examples in which the behavior of a time-critical processing needs to be periodically monitored and adapted throughout the execution.

One of the most widespread programming models for parallel applications is based on the **MPI** (Snir et al., 1995) (Message Passing Interface) communication library. In this case parallel programs are expressed in an un-structured way by describing the behavior of a set of distributed processes cooperating by using communication channels. Last implementations of the MPI library provide some form of support to dynamic reconfigurations. MPI2 (Lusk, 2002) provides a mechanism for instantiating new processes at run-time: this feature can be exploited in such a way as to perform structural changes of a parallel program, for instance it is possible to increase the parallelism degree (e.g. the number of executed processes) achieving in this way an expected better performance. In general the management of adaptivity issues in MPI is completely left to the programmer, which is heavily involved in ensuring and maintaining the consistency and the correctness of the computation during and after the reconfiguration phase.

In order to partially solve the complexity issues of un-structured parallel programming, some notable research works have been proposed as the **ASSIST** (Vanneschi, 2002) programming environment. ASSIST is a general parallel programming framework for several classes of computing architectures, from shared-memory platforms as SMP and NUMA multi-processors to distributed-memory multicomputers as cluster of workstations and large-scale Grids (Coppola et al., 2007a). The most important novelty of this approach is the structured methodology for expressing parallel computations, which are instances of well-known parallelism schemes (e.g. task-parallel programs as task-farm or pipeline and data-parallel computations as map, reduce or communication stencils). This approach is known to the scientific community as *Structured Parallel Programming* (Cole, 2004).

In ASSIST a run-time support to dynamic reconfigurations (Aldinucci et al., 2005; Vanneschi & Veraldi, 2007) is rendered by exploiting a transparent-approach to adaptivity. A dynamic reconfiguration is a change involving a specific application component by modifying: (i) the mapping between execution processes and underlying computing resources (i.e. geometrical changes); (ii) by increasing or decreasing the number of execution processes of a component (i.e. structural changes). Some papers (Coppola et al., 2007b; Danelutto et al., 2007) describe how adaptivity is exploited in the ASSIST framework. Reconfigurations are *performance-oriented* (Aldinucci et al., 2006): dynamic changes of application components are triggered by the run-time support in presence of QoS violations of a pre-defined performance contract.

Another interesting work is the **Behavioural Skeleton** (Aldinucci et al., 2008) approach. Adaptivity for distributed high-performance computations is exploited by means of the Grid Component Model (Ahumada et al., 2007; Mathias et al., 2008) (GCM) and the structured parallel programming paradigm (which the authors also call skeleton-based programming). In GCM a self-adaptive component is composed of two main parts: the Membrane which is an abstract unit responsible for controlling the adaptive behavior of the component, and the Content composed of a set of processes performing the corresponding functional logic of the computation. These entities can also be other GCM components (i.e. inner components): therefore the GCM model makes it possible a natural hierarchical nesting between several self-adaptive components.

This approach is characterized by very interesting run-time mechanisms for controlling multiple non-functional concerns (e.g. it is possible to simultaneously control different parameters as performance and security objectives). In this case the solution proposed in (Aldinucci et al., 2009) provides multiple autonomic managers for a single component, each one controlling a specific non-functional concern by using a set of event-condition-action (ECA) policy rules. Different policies can lead to some conflicting decisions: in this case the authors propose distributed consensus-based solutions.

In this section we have presented the actual state of the art concerning self-adaptive systems for mobile applications and for traditional high-performance computing problems. From our point of view there is not yet a unified approach for programming adaptive high-performance computations executed on highly heterogeneous and dynamic execution environments, such as pervasive and mobile grid infrastructures. Some research works focus on HPC computations in real-time environments, but in these approaches the "mobile part" of application definition is essentially missing. Other research works achieve the necessary expressiveness to define mobile adaptive applications, but they do not face on compute-intensive real-time computations.

3. Application scenarios

In this chapter we are interested in describing the design and the developing issues of time-critical systems as emergency management applications. They are a notable example of systems that exploit highly heterogeneous distributed computing platforms, composed of traditional high-performance architectures as well as mobile nodes and sensors, providing system features and services according to precise QoS constraints. Such QoS is typically related to the performance at which computing results or communication facilities are provided to users and/or to the service availability and reliability w.r.t. software and hardware failures.

A main issue in executing this class of applications on heterogeneous platforms is hence given by the chance of defining proper programming models and run-time supports aiming at enabling the definition and dynamic satisfaction of QoS constraints. This issue can be seen as even more complex by considering the strong dynamic variability of computing and communication services provided by these platforms, also given by the mobility of some of its nodes and their geographical distribution.

Emergency management systems include data- and compute-intensive processing e.g. forecasting and decision support models) not only for off-line centralized activities, but also for on-line and decentralized activities. Consider the execution of software components performing a forecasting model, which is a critical compute-intensive computation to be executed respecting precise operational real-time deadlines. In "normal" conditions (e.g. concerning the network availability and the emergency scenario itself), we could be able to execute these components on a centralized server, exploiting its processing power to achieve the highest performance possible. Critical conditions in the application scenario (e.g. an emergency detection) can lead to different user requirements (e.g. increasing the performance to complete the forecasting computation within a given, new deadline). Moreover, changes in network conditions (e.g. of the interconnection network with HPC centralized resources) or in computing resource availability can lead to the necessity to execute proper versions of the application components directly on localized mobile resources which are available to the users (e.g. civil protection personnel, rescuers and stakeholders).

In such cases, compute-intensive elaborations that need to be periodically executed for monitoring, predicting and taking response actions during an emergency phase, can be alternatively executed on different or additional computing resources, including sets of distributed mobile resources running properly developed versions of these computations. In other words, in these scenarios it is important to assure the *service continuity*, adapting the application to different user requirements but also to the so-called *execution context*, which corresponds to the actual conditions of the both the surrounding environment and the computing and communication platform. So the key issues in the definition of high-performance parallel programming paradigms, models, and frameworks to design and develop these kinds of complex and dynamic applications, is adaptivity (possibility to adapt the application behavior changing component versions and the platforms on which they are executed) and **Context Awareness** (knowledge about the current condition of the reference environment).

The general reference point for these kinds of applications is the Grid paradigm (Berman et al., 2003) which, by definition, aims to enabling the access, selection and aggregation of a variety of distributed and heterogeneous resources and services. However, though notable advancements have been achieved in recent years, current Grid technology is not yet able to supply the needed software tools to match the high-performance feature with high adaptivity, ubiquity, proactivity, self-organization, scalability, interoperability, as well as fault tolerance and security, of the emerging applications running on a very large number of fixed and mobile nodes connected by various kinds of networks.

We claim that a *high-level programming model* is the only viable solution to design and develop such kind of distributed applications. A novel approach needs to provide a proper combination of high-performance programming models and pervasive and mobile computing frameworks, in such a way to express a QoS-driven adaptive behavior for critical high-performance applications on heterogeneous contexts. In the following section the main features of a novel approach will be discussed in more detail.

4. Features and requirements of a novel approach

A novel approach needs to be characterized by a strong synergy between two different research fields: *Pervasive and Mobile Computing* (Hansmann et al., 2003) and *Grid Computing* (Berman et al., 2003). Pervasive and Mobile Computing is centered upon the creation of systems characterized by a multitude of different computing and communication resources, whose integration aims at offering seamless services to the users according to their current and time-varying needs and intentions. On the other hand Grid Computing focuses on the efficient execution of compute-intensive processes by using geographically distributed sets of computing resources.

To merge these two areas, an effective integration must provide a proper combination of high-performance programming models and mobile computing frameworks in such a way as to express a QoS-driven adaptive behavior for distributed applications like emergency management systems. In this section we will highlight the main guidelines that need to be followed for exploiting such integration. They can be summarized in three different points: (i) how compute-intensive processing for heterogeneous environments can be programmed and their performance formally analyzed; (ii) how a distributed application can change its QoS behavior at run-time; (iii) when it is necessary to execute an application reconfiguration.

4.1 Structured Parallel Programming

First of all we require a structured methodology for expressing parallel and distributed programs in such a way as to make the programming effort less costly and less time-consuming and to have a predictable QoS behavior for such applications. This means that we need sufficient high-level abstractions featuring a large-degree of programmability, compositionality and *performance portability*. Portability plays a central role in heterogeneous contexts: portable parallel applications are able to achieve acceptable performance results on different computing platforms, trying to exploit in the best way as possible the underlying physical aspects of the target machine. This can be done by tuning proper application parameters (e.g. parallelism degree and task granularity) statically, during the compiling phase, or dynamically through run-time reconfiguration processes.

For these reasons *Structured Parallel Programming* (SPP) (Cole, 2004) has been proposed as an effective and attractive approach to parallel programming featuring interesting properties in terms of high-level programmability and performance portability. According to the SPP methodology a parallel computation is expressed by using well-known abstract parallelism schemes for which parametric implementations of communication and computation patterns are known. In fact the experience in parallel programming suggests that parallel programs make use of a limited number of parallelism patterns exhibiting regular structures, both concerning data organization and partitioning or replication of functions. In this way we can identify several parallelism paradigms as *data parallelism* schemes (e.g. map, reduce, parallel prefix and communication stencils) and *task parallelism* structures (e.g. pipeline, task-farm and data-flow). Furthermore the performance behavior of these parallelism schemes has been studied by exploiting formal analysis (i.e. *performance models*) based on queueing theory and queueing network fundamental results, thus rendering the performance modeling of this class of computations usable also from automatic tools as compilers (e.g. for statically deciding the best application configuration) and from run-time supports (e.g. for providing efficient fault-tolerance (Bertolli, 2009) mechanisms but also for deciding the execution of dynamic reconfigurations (Vanneschi & Veraldi, 2007)).

The SPP methodology is based on the concept of parallelism schemes that exhibit the following features: (i) they are characterized by constrains in the parallel computation structure; (ii) they have a precise semantics; (iii) they are characterized by a specific performance model; (iv) they can be composed each other to form complex graph computations. A central property of these schemes is the complete knowledge of their computation structure, e.g. in terms of data and functions replication and/or partitioning. For instance task-farm paradigm exploits replication only, applied to functions and to non-modifiable data. Pipeline exploits a partition of the sequential elaboration in a sequence of multiple successive phases each one using a set of replicated (non-modifiable) or partitioned (modifiable) data. On the opposite direction data-parallel schemes correspond to the replication of the same functionality and to the partitioning of data, so that distinct parallel units are able to apply the same operations to different data partitions in parallel.

Performance of parallel computations can be measured in terms of throughput, i.e. completed tasks per time unit, and computation latency, i.e. the average time needed to execute the computation on just one input task. Parallelism schemes can have different impacts on these two performance measurements. Task-farm and pipeline, though able to increase the throughput of the computation, also tend to increase the latency compared to the sequential implementation due to communication overhead. Data-parallel and data-flow tend to improve both the computation latency and the throughput, but they can be hampered by load-balancing issues. Finally parallelism schemes are also characterized by a different degree of memory utilization. Task-farm tends to increase the memory capacity, the others, being based on data partitioning, tend to decrease it.

As far as composability is concerned, parallel distributed applications can be represented as graphs of modules. Each module can exploit a sequential or a parallel computation based on a specific parallelism scheme (*intra-module parallelism*), whereas modules can be composed in general complex graph structures (*inter-module parallelism*).

4.2 Dynamic reconfigurations of parallel programs

Previously in this chapter we have introduced the concept of application configuration. Any run-time activity that modifies the parallelism degree, the parallel version (e.g. parallelism scheme and parallelized sequential algorithm) or the execution platform of a module is indicated as a *dynamic reconfiguration process*.

Although there are specific situations in which the "best" configuration for each module can be statically selected during the system design phase (e.g optimizing performance parameters and memory and energy usage), in dynamic execution conditions this best configuration might not be identifiable without run-time information. As an example consider the following situations:

- in dynamic environments, the degree of availability of computing and network resources can dynamically vary also in unexpected ways;
- for compute-intensive elaborations (as forecasting models and simulations), precise user requirements can be requested. For instance when an emergency scenario is detected, the decision support model that assists civil protection personnel could require to be executed until a maximum completion time;
- for irregular parallel problems, characterized by computations whose size, distribution and complexity depends on the properties of input data, a run-time support to dynamic reconfigurations is an unavoidable feature that needs to be provided.

For structured parallel computations we can classify the set of adaptation processes involving a single application module in two categories namely non-functional and functional reconfigurations.

Non-functional Reconfigurations are adaptation processes involving the run-time modification of some implementation aspects of a parallel module of the application graph. For implementation aspects we intend that:

- it is possible to modify the current parallelism degree of a module, e.g. increasing the number of implementation processes in such a way as to achieve a better performance level (e.g. response time for each request or the completion time of the entire computation);

- the run-time support can modify the mapping (i.e. geometrical changes) between implementation processes of a parallel module and underlying physical computing resources on which they are executed;

- the run-time support can provide another important class of geometrical changes involving the run-time *data re-distribution* among the set of worker processes of a parallel module. Such changes are an effective approach for addressing potential workload un-balancing problems of data-parallel parallelizations.

The most important aspect characterizing non-functional reconfigurations is that in every case these processes do not modify the sequential algorithm performed by the parallel module, neither the exploited parallelization scheme.

Another important and, in some sense, intuitive concept is that the same problem can usually be solved in a parallel fashion by exploiting several different parallelization schemes. For instance let us consider a simple example in which a parallel module periodically receives a sequence of input tasks each one consisting in a pair of a matrix and a vector. For each task the module performs a matrix vector product and the result vector is transmitted to a destination module. According to the way in which we perform data distribution among parallel entities, we can parallelize this problem in several ways. A task-farm scheme consists in replicating the entire matrix-vector computation on multiple independent workers that apply the computation on a different set of scheduled input tasks. In this way we are able to improve the throughput of the computation but not the computation latency. For this simple example several data-parallel schemes can also be adopted. For each input task, if we partition the input matrix among a set of parallel processes and we replicate the input vector, we are able to exploit a classic data-parallel *map* scheme in which each process exploits the matrix-vector computation on its partition without requiring inter-process data exchange and synchronizations. Moreover, if the input vector instead of being replicated is partitioned among processes, we are able to exploit a data-parallel approach with a *communication stencil*. In both the cases the data-parallel approach improve latency of the computation as well as its memory occupation w.r.t the task-farm implementation. Moreover the stencil version requires a higher number of exchanged data during the computation.

The previous scenario suggests that we can use this important feature of structured parallel programming in such a way as to effectively deal with highly dynamic and heterogeneous execution environments. For each application module several alternative versions can be provided, and **Functional Reconfigurations** are implementation changes that consist in changing at run-time the version which is currently executed by a parallel module. Versions can have a different but compatible semantics: they can exploit different sequential algorithms, parallelization schemes or optimizations, but preserving the module interfaces in such a way that the selection of a specific version does not modify the global application behavior. Each version of the same parallel module is optimized for being executed whenever certain execution conditions are satisfied (e.g. depending on the actual levels of communication bandwidth, power supply, available memory) or on the presence of specific classes of computing resources.

Another important advantage of this methodology w.r.t non-structured parallel programming models is that it renders feasible a transparent approach to application reconfigurations (see Section 2). As demonstrated in (Aldinucci et al., 2006; Vanneschi, 2002; Vanneschi

& Veraldi, 2007), the reconfiguration code that performs non-functional reconfigurations (as parallelism degree variations) can be automatically extracted from the knowledge of the parallelism scheme that needs to be modified. In this case the programmer is only involved in providing the implementation of the high-level functions executed by a set of worker processes, and their number and the interconnections for task reception and result transmission are completely managed by the run-time support system. Moreover, for functional reconfigurations, in (Bertolli et al., 2011) formal consistent reconfiguration protocols have been presented for switching between different structured parallelism schemes.

4.3 Exploiting adaptation strategies for parallel programs

An adaptive application is a system which evolves over time changing its initial configuration in response to external events. Providing a run-time support to dynamic reconfiguration activities is a first and essential requirement for developing adaptive high-performance applications. Although many research works (Aldinucci et al., 2008; 2005; 2006; Coppola et al., 2007a; Vanneschi & Veraldi, 2007) focus on specific implementation issues describing several optimizations for reducing reconfiguration costs in terms of performance degradations for certain parallelization schemes and their compositions, they do not pay sufficient attention to the decision process that triggers the execution of these reconfigurations. We refer to this process as an *Adaptation Strategy*, that is a plan or a set of rules used by the system for achieving the specific objectives of its adaptive behavior.

Especially for emergency management systems several constraints need to be satisfied throughout the execution. The system must offer its functionalities to the users according to different QoS notions:

- in many adaptive systems we need to control the execution progress preserving observed metrics within user-defined ranges. Classic examples are: e.g. maintaining the mean response time of the system within a *window of tolerance*, i.e. between a maximum and a minimum acceptable threshold over the execution. In this case we refer to this objective as a **threshold specification** problem;

- alternatively we might require to maintain some quantitative execution metrics as closer as possible to a set of desired reference values. For instance it could be necessary to maintain the mean service time of a computation equal to a specified target value desired by the final user. We refer to this approach as a **set-point regulation** problem;

- more often we need to control several execution measurements of the system, as the performance, energy and memory requirements, number of utilized computing and network resources. In this situation we need to find a control law such that a certain optimality criterion is achieved. This problem can be casted into a mathematical fashion introducing an objective function $F(x_1, x_2, \ldots, x_n)$ of n input parameters whose value should be minimized or maximized. Inputs are system controllable features, and the adaptation strategy is aimed at optimizing this multi-variable function during the system execution. This formulation is known as an **utility/cost optimization** problem.

In different research fields (e.g. Autonomic Computing (Huebscher & McCann, 2008) and Control Theory) general adaptive systems have been described in terms of a reference control architecture (see Figure 1) composed of two main parts and proper interaction phases between them. The two parts are:

- the target controlled system (namely *plant*) that we want to control. It takes *functional inputs* and generates *functional outputs* according to the semantics of its computation. However, for controlling the behavior of a system, other relationships among non-functional inputs and outputs are also important. System execution can be influenced by a specific set of manipulated *control inputs*, whose values strictly affect the way in which the system execution is exploited. Based on the desired control objectives, the quality of system operation is evaluated by measuring non-functional outputs also called *observed outputs* (e.g. the current level of some performance metrics);
- a *controller*, which is an independent entity able to monitor and affect the system operational conditions by taking and analyzing observed outputs and promptly decide the corresponding set of control inputs.

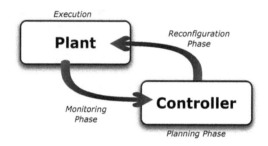

Fig. 1. Control-loop scheme of an Adaptive system: the cycle evolves according to a monitoring phase, a planning phase and a reconfiguration phase.

Interactions between these entities follows a classic closed-loop feedback scheme composed of different phases. *Monitoring phase* involves capturing current properties and measurements of the system which are effective for identifying when the execution of a dynamic reconfiguration can be useful for achieving the desired objectives. *Planning phase* consists in a set of concrete actions aimed to select a new set of control inputs that are the best response to the current observed outputs of the system. Finally the *Reconfiguration phase* applies the decided set of control inputs to the controlled system. In this part of the chapter we are interested in understanding how the planning phase can be exploited and what methodologies can be used. The following two sections will present two different approaches that can be used for addressing the adaptation strategy specification for adaptive systems.

4.3.1 Adaptation strategies expressed by logic rules

A flexible methodology for expressing adaptation strategies, which has been exploited in Pervasive and Mobile scenarios due to its intrinsic simplicity and programmability, consists in defining adaptation processes as reactions to specific system situations. As an example if the available energy level of a mobile device is lower than a specified threshold (e.g. 25 %), the execution of a mobile application can switch to a low-energy version which preserves the battery duration employing a limited utilization of the device display. Therefore a straightforward solution for expressing adaptation strategies consists in providing a mapping between events and execution situations and corresponding reconfigurations (i.e. situation-action pairs) as a finite set of imperative logic rules. System administrators and designers encode management guidelines as a set of rules that represent the adaptation policy of the system.

Policy rules are a form of guidance used to determine decisions and corresponding actions on the system execution. They have been introduced especially in the field of intelligent agents Russell & Norvig (2003), where abstract entities are able to perceive their environmental conditions and act on the basis of this information in order to maximize their objectives. In this vision an adaptive system is in a specific internal state at every given moment of time, and a set of logic rules may cause an action to be taken an therefore a transition to a different internal state of the system.

Logic rules can be expressed according to several paradigms. Although many techniques exist, with specific modifications and features depending on the particular application context, a general approach is based on **Event-Condition-Action** (ECA) policy rules for programming reactive behaviors. ECA rules have the following basic syntax: _when_ event _if condition then action;_. Informally the abstract semantics is: the occurrence of the event triggers the rule evaluation, the condition is checked in order to ensure that the system is in a specific internal state. If this condition holds, the corresponding action is enforced. Notice that though the state that will be reached by applying the rule is not explicitly expressed, the policy programmer knows the desired effect of the selected action.

If multiple rules are fired simultaneously, the entire set of corresponding actions will be performed at the same time. This may lead to potential conflicting actions (e.g. resulting in conflicting post-adaptation situations). The design of proper techniques for efficiently identifying and resolve such conflicts has been an intensively studied research issue. Although the large research effort, the general problem of avoiding conflicting rules is hard to be solved, especially for complex systems in which the consequence and the side-effects of a dynamic reconfiguration involve the long-term steady-state behavior of the system (i.e. in this case a rule can conflict with rules that may be triggered on different events in future). This suggests that, although highly programmable and user-oriented, _adaptation strategies based on logic rules are difficult to be applied to complex applications as emergency management systems_. In fact in this case it is extremely important to have a reasonable expectation about the consequences of reconfiguration actions on the long-term behavior of the systems (e.g. future performance level after the completion of a reconfiguration process), that can be difficult to be evaluated and represented through situation-action pairs.

4.3.2 Adaptation strategies based on control-theoretic techniques

Besides Artificial Intelligence and Reactive Systems, the concept of automated operations is an important research field in disciplines as mechanical and electrical engineering for developing autonomous systems able to respond to changing workloads and expected run-time conditions. The methodology based on Control Theory (Hellerstein et al., 2004) foundations has been intensively exploited for design controllers and feedback systems in many industrial plants and mechanical infrastructures. These solutions provide powerful mechanisms for dealing with unpredictable changes, uncertainties, and system disturbances.

In many research works several control-theoretic modelings of computing and networking systems have been proposed. Control-based approaches are often a solid solution to solve network problems like congestion and flow control, and rate adaptation of queueing networks. This leads to an increasingly important research area in which the adaptive system is the computing network itself, i.e. Autonomic Networking (Mortier & Kiciman, 2006), whose interconnection facilities are able to automatically detect, diagnose and repair failures, as well as to adapt the underlying network configuration and optimize its performance and quality of service.

Nevertheless the exploitation of control theory methodologies to computing systems is rarely used in practise. The formal design of controllers needs the ability to precisely quantify the reconfiguration effects on the system evolution. Hence, *system modeling* is a prerequisite for applying such techniques. This model needs to be expressed in an input-output form concerning how control inputs and disturbances affect the behavior of observed measurements. Such relationships between model variables can be determined through *first-principle models*, based on the actual physics laws which govern the evolution of the system, or exploiting *empirical models* in which the relationships are extracted through static techniques (e.g. least-square techniques) on properly defined experimental data and observations.

We claim that a relevant example in which first-principle models can be used consists in the performance modeling of structured parallel computations. For these structured parallelization schemes and their composition in computation graphs, performance models provide a quite acceptable performance predictions of system steady-state behavior, which is often a sufficient guideline to define powerful control-theoretic strategies for controlling parallel applications. When it is possible, these approaches are amenable to provide properties like the *accuracy* of an adaptation strategy (how precisely control objectives are satisfied), *settling time* (how long a reconfiguration lasts until the steady-state behavior of the system is reached) and the *stability degree* of a reconfiguration (how long a system configuration represent a "good choice" for achieving the desired system objectives).

For these reasons control theory techniques, though their applicability is not always straightforward, represent a valuable research direction for controlling time-critical applications as emergency management systems. Based on the main guidelines described in this section, in the rest of this chapter we will introduce in more detail our approach.

5. Design principles of a novel programming model approach

As stated in Section 4.1 our approach is based on a structured methodology in which distributed and parallel applications can be represented as directed graphs of modules interconnected through streams of data [1]. Whereas the structure of the graph can be arbitrary (e.g. cyclic client-server interactions or acyclic graphs of multiple computation phases), the internal parallelism inside each module is expressed by means of structured parallelism schemes.

The adaptive behavior is expressed in the possibility of each parallel module to perform dynamic reconfiguration activities triggered by their own *control logic*. This means that adaptivity is exploited through a set of independent entities that cooperate for implementing the distributed *functional logic* of the application and also for negotiating reconfigurations

---

[1] A stream of data is a sequence, possibly of unlimited length, of typed data structures.

taken by their control logics. In this section we will introduce the basic structure and the formalization of the concept of adaptive parallel module.

5.1 Structure and modeling of an adaptive parallel module

The core element of our approach is the concept of adaptive parallel module (shortly **ParMod**), an independent and active unit featuring a parallel elaboration and an adaptation strategy for responding to different sources of dynamicity. From an abstract point of view a ParMod can be structured in two interconnected parts:

- an **Operating Part** is responsible for performing a parallel computation expressed according to a certain parallelism scheme of the SPP framework (e.g. task-farm and data-parallel are relevant examples). Without loss of generality we assume a stream-based computation in which a set of input data streams from other parallel modules are received by the operating part. The parallel elaboration is activated according to a non-deterministic selection of input elements or, based on a data-flow semantics, waiting for the reception of an element from each input interfaces of the module. Result externalization (if it is necessary) is exploited onto output data streams (output interfaces) to other parallel modules;
- a **Control Part** is an autonomous entity able to observe the operating part execution and modify its behavior exploiting reconfiguration activities.

For each ParMod we suppose the presence of multiple alternative configurations of its operating part that differ in the parallelism degree, the structured parallelism scheme or in the execution platform on which they can be deployed. The only constraint that we impose is that every configuration must respect the same input and output interfaces of the ParMod (i.e. admissible reconfigurations are geometrical, structural or implementation changes, see Section 2), in such a way that a local reconfiguration involving a single parallel module does not modify the interfaces provided to other application modules. Therefore each ParMod features a multi-modal behavior defined as follows:

Definition 5.1. (Multi-modal behavior of ParMod Operating Part). At each point of time the operating part can behave according to a certain active configuration belonging to a specific set C of alternatives:

$$C = \{C_0, C_1, \ldots, C_{v-1}\} \tag{1}$$

This set represents a *finite and discrete set of statically known alternative configurations* of the parallel module.

The adaptive behavior of a parallel module is exploited through a periodical information exchange between operating and control part. *Observed outputs* are QoS measurements (e.g. memory occupation, energy consumption, number of completed tasks) that are periodically monitored by the control part. *Control inputs* are proper commands that modify the current operating part configuration exploiting non-functional or functional reconfiguration activities. Observed outputs and control inputs exchange can be exploited periodically, at fixed points equally spaced (in this case we speak about a *time-driven* controller) or whenever specified events are satisfied (as in the case of *event-driven* controllers).

Our ParMod control model is based on a time-driven control part that takes reconfiguration decisions every **control step** of duration τ. In other words at the beginning of each control step the control part acquires observed outputs from operating part, executes a specific control algorithm to decide a corresponding set of control inputs that will be communicated to the operating part.

5.1.1 A Hybrid modeling of the operating part behavior

In order to apply a formal methodology for developing ParMod adaptation strategies, we need to model the QoS temporal evolution of a parallel computation by using a mathematical and formal approach. This means that we need proper mathematical equations that express the relationship among observed output measurements and current control inputs. As we have seen in Section 4.3.2, the structured parallel programming paradigm is a basic point in order to establish such modeling. Relevant examples are:

- structured parallelism schemes feature a sort of predictability of their steady-state performance level. Fixed the parallelism degree and the execution platform, several performance measurements as the mean service time, the mean queue length and the computation latency can be predicted and formally analyzed according to analytical models based on queueing theory and queueing network results (Vanneschi, 2002);

- memory utilization models can also be derived for well-known parallelism schemes by exploiting the specific behavior of different parallelization patterns in terms of function and data replication or partitioning (Bertolli et al., 2010);

- structured parallelism schemes have a precise semantics in terms of computations performed by each parallel unit, the size of exchanged messages and the frequency of certain activities as calculation and message transmission. Such knowledge can be a starting point in order to define models for measuring the power consumption of a parallel module, especially when its execution is mapped onto energy-limited resources as mobile devices.

A formal model that describes the predictability of QoS parameters involves the following set of variables:

- *state variables* represent system measurements that are useful to maintain during successive control steps. These variables can describe the global energy consumption caused by the parallel execution of a ParMod, the number of queued tasks, the global number of completed input stream elements. We refer with the term $\mathbf{x}(k)$ as the current value of the internal state of the operating part model at the beginning of the control step k. State variables are directly monitored at each control step by the control part;

- *disturbance variables* are uncontrolled exogenous signals that can affect the relationship between control inputs and the observed state variables modeling the actual QoS of the computation. Uncontrolled means that the control part is not able to decide and fix their values, but they are determined by environmental or external decisions outside of ParMod control. Relevant examples are the mean calculation time of a sequential function on a certain target architecture, the communication latency and the mean inter-arrival time of requests to the system. We refer with the term $\mathbf{d}(k)$ as the value assumed by disturbances throughout the k-th control step;

- *control inputs*, $\mathbf{u}(k)$, indicate the ParMod configuration that will be exploited throughout the k-th control step.

According to the multi-modal behavior of the operating part and the QoS predictability of different ParMod configurations, we can identify two classes of transitions that characterize the ParMod execution:

- *continuous transitions*: when a configuration has been fixed, the evolution of continuous-valued QoS state parameters can be predicted by applying a specific mathematical model corresponding to the currently used configuration;
- *discrete transitions*: by executing reconfiguration activities, the current configuration can be changed passing from a configuration C_i to a different alternative configuration C_j, shortly $C_i \rightarrow C_j$ with $i \neq j$.

The presence of continuous transitions (of continuous variables) and discrete transitions (of alternative configurations) suggests to model the operating part of a ParMod as an *hybrid system* (der & Schumacher, 1999) in which these two dynamics are formally modeled in a unique and refined mathematical structure.

For each configuration $C_i \in C$ the temporal evolution of observed outputs and of the next state variables can be expressed by a model in a state-space form as shown in (2).

$$x(k+1) = \phi_i\big(x(k), d(k)\big) \tag{2}$$

The semantics of the next state expression provides that, if the values of expected measured disturbances and the current model state for control step k are known, by applying a function ϕ_i we can predict the values assumed by state variables at the beginning of the next control step $k + 1$.

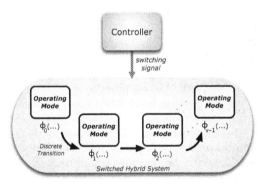

Fig. 2. The operating part evolves according to continuous transitions (i.e. concerning QoS variables predicted through the model of the current configuration), and discrete transitions (reconfigurations).

Two considerations are important at this point. First of all the function ϕ_i can be obtained by applying statistical methods or, as we have seen, by using a higher level knowledge of the structure of the computation behavior in order to extract first-principle relations. The second point is that each model ϕ_i is strictly coupled with a configuration of the ParMod. In some cases a unique mathematical model can be parametrically provided for multiple configurations (e.g. as for different parallelism degrees), but there are also situations in which configurations can require a completely different modeling of the state variables dynamics (e.g. as for the case of parallel versions expressed by means of different parallelism schemes as task-farm and data-parallel ones).

This variable-structure behavior of the operating part is typical of a large class of hybrid systems featuring a limited set of alternative operating modes (i.e. configurations in our case). Such class consists in the so-called *Switched Hybrid Systems* Liberzon (2003) (see Figure 2) in which control inputs from the controller modify the actual configuration of the operating part, and future QoS measurements can be predicted following the evolution rule strictly coupled with the new configuration. In conclusion the entire operating part model is given by:

$$
\begin{aligned}
\mathbf{x}(k+1) &= \Phi\Big(\mathbf{x}(k), \mathbf{d}(k), \mathbf{u}(k)\Big) \\
&= \text{if } \Big(\pi(\mathbf{u}(k)) = C_i\Big) \text{ then } \phi_i(\mathbf{x}(k), \mathbf{d}(k))
\end{aligned}
\tag{3}
$$

in which function Φ is a piecewise-defined function that provides a different definition for each alternative configuration of the parallel module, whereas π is a bijective function that maps each possible value of control inputs onto corresponding configuration indices.

5.2 A Control-theoretic adaptation strategy for ParMods

In this section we will propose the exploitation of a control-theoretic technique for controlling the QoS behavior of an adaptive parallel module. We point out that in the context of time-critical applications, like emergency management systems, a classic reactive approach in which the system reacts to well-identified circumstances in pre-programmed way (as in classic environments for adaptive HPC applications) is not an effective adaptation strategy. The criticity of QoS requirements for these applications imposes that reconfiguration decisions must be selected taking into account the history of the system, trying to respond in a pro-active fashion to the user requirements and to the future behavior of the surrounding execution environments. In this case we need adaptation strategies that are more similar to the ones studied in pervasive systems (see Section 2), in which taking actions in advance to potentially critical future events is a central point of this approach.

The *predictive control* approach that we propose for adaptive parallel applications is based on the following points:

- the presence of (empirical or first-principle) mathematical models that can be used to estimate future QoS behavior of the computation in function of a planned sequence of control inputs (i.e. a *reconfiguration plane*) and future predictions of disturbances;
- the presence of an objective function describing the control aims which drive the selection of the optimal control input sequence;
- the specification of boundary conditions on state and control input variables.

This formulation of the predictive control problem is known to control theorists as *Optimal Control* (Bertsekas, 1995). Optimal control is the process of determining control and state trajectories for a system over a certain period of time in order to optimize a properly defined objective function. A typical representation of an objective function is given below:

$$
\max \ U(k) = \sum_{i=k}^{k+h-1} L\Big(\mathbf{x}(i+1), \mathbf{u}(i)\Big)
\tag{4}
$$

The function represents an aggregate utility (or cost) which depends on the desirability of future system internal states and control inputs taken for a horizon of N successive control steps (e.g. ideally the whole execution duration). Expressing control preferences among different future states and control input trajectories is an expressive way to define powerful control strategies for structured parallel computations featuring a predictable behavior. Objective functions can be tuned to satisfy the specifications of the system, the control objectives as well as to find feasible trade-offs between conflicting requirements (e.g. optimization problems can exploit an intrinsic multi-variable nature including performance, power and memory measurements and specific constraints on state and control variables).

If we suppose to know the exact trajectory assumed by disturbance inputs for the whole execution duration, and if the system model is sufficiently accurate and precise, we can statically define an optimal reconfiguration plane that optimizes the ParMod execution. Of course this assumption is not always feasible for the following reasons:

- disturbance inputs are variables whose behavior (e.g. average values) can not be statically known a priori but they may depend on uncontrollable factors (as the actual conditions of the underlying execution platform);
- the system model can be effected by perturbations and unmodeled dynamics (e.g. due to unmeasured disturbances) that limit the quality of the future QoS estimations.

For the previous reasons applying the optimal reconfiguration plane step-by-step in an open-loop fashion is not a viable approach. Nevertheless several sub-optimal approaches are available in order to iteratively apply optimization problem solutions in a closed-loop fashion. In the next section one of this technique, namely the *Model-based Predictive Control*, will be introduced.

5.2.1 Model-based predictive control of structured parallel computations

Model-based predictive control (Garcia et al., 1989) (shortly **MPC**) is a repetitive procedure that combines the advantages of long-term planning (feedforward control based on system predictions over a future horizon) with the advantages of a feedback control using actual system and disturbance measurements. At the beginning of each control step k the values of the model state variables and of past disturbances are measured by the control part. At this point a limited future time horizon (i.e. a *prediction horizon*) of h consecutive control steps is considered, and a prediction of disturbances for this short time interval is exploited through proper statistical techniques. The predicted disturbance trajectory is composed by:

$$D_k^{k+h-1} = \{\hat{\mathbf{d}}(k|k), \hat{\mathbf{d}}(k+1|k), \ldots, \hat{\mathbf{d}}(k+h-1|k)\} \tag{5}$$

where the syntax $\hat{\mathbf{d}}(k+i|k)$ means that the estimated value of disturbance inputs for the step $k+i$ is predicted using the current knowledge at control step k.

These predictions are used to plan a sequence of optimal reconfiguration decisions for each control step of the prediction horizon. The optimization problem (4) is solved for the limited time horizon finding an optimal control input trajectory:

$$U_k^{k+h-1} = \{\mathbf{u}(k|k), \mathbf{u}(k+1|k), \ldots, \mathbf{u}(k+h-1|k)\} \tag{6}$$

Instead of applying the optimal reconfiguration plane step-by-step, the uncertainty of disturbance estimation (which increases going deeper in the prediction horizon) and the potential inaccuracy of the future QoS predictions, suggest a more effective approach based on an iterative procedure. Only the first control decision of the optimal trajectory is transmitted from the control part to the operating part at the beginning of the current step k while the rest are discarded. This process is repeated at the beginning of the next control step when: (i) the new system state is available; (ii) values of past disturbances for the previous step have been measured. The effect of this control algorithm is to move the prediction horizon towards the future following the so-called *receding or rolling horizon* technique.

Based on the QoS predictability of structured parallelism schemes, the benefits of the predictive approach are evident. Suppose to have a ParMod control objective formulated as a threshold specification problem: e.g. we need to maintain the observed mean throughput (completed tasks per time unit) of a parallel computation within an acceptable region of values (e.g. between a maximum and a minimum threshold established by the user). Although an initial configuration (e.g. parallelism degree) can be identified during the design phase such that this QoS requirement is met, future modifications of the initial conditions modeled as disturbances (e.g. time-varying task grain) can prevent this configuration to be no longer acceptable. Especially for time-critical applications the utilization of a predictive strategy can be a valuable solution for two main reasons:

- selecting in advance reconfiguration actions can be crucial in order to anticipate undesired behaviors and promptly mitigate the future variability of disturbances;
- typically reconfigurations involve a cost on the execution. After a reconfiguration a ParMod can incur some dead-time, e.g. for completing reconfiguration protocols (see (Bertolli et al., 2011)) or rather the cost for changing the number of currently used computational resources (e.g. as in Cloud Computing environments).

Despite MPC is a largely used approach in many real-world scenarios (e.g. for controlling chemical, mechanical and industrial plants), it is considered a compute-intensive technique especially for hybrid systems (as a ParMod) where the set of control inputs is discrete. In this case the on-line optimization problem requires to explore a search space whose size grows exponentially with the length of the prediction horizon. The search space can be represented as a complete tree (i.e. the ParMod *evolution tree*) with a depth equal to the prediction horizon length h and an arity that coincides with the number of ParMod configurations v. The number of explored nodes is given by $\sum_{i=0}^{h} v^i$ thus exponential in the prediction horizon length. If not properly addressed with specific techniques (e.g. branch&bound approaches) or heuristics, the combinatorial optimization problem that needs to be solved at each control step implies an exhaustive search by testing among all the feasible combinations of reconfigurations, thus potentially limiting its viability to systems with long sampling intervals. Nevertheless in practical scenarios, since the number of ParMod configurations is often sufficiently limited and prediction horizons are normally short due to disturbance prediction errors, this approach can be exploited in practise without complex search space reduction techniques.

6. Experiments discussion

In this section we will report a summary of experimental results that we have presented in a more comprehensive way in our past publications. Our test-bed application describes a limited but interesting part of a complex emergency management system for flood predictions, that we have already introduced in Section 3. The application graph is depicted in Figure 3. The system is responsible for providing short-term forecasting results of a flood

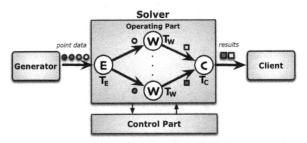

Fig. 3. Flood Emergency Management Application: this test-bed scenario considers a limited set of application modules.

scenario in a real-time fashion to a set of interested users (e.g. emergency stakeholders or rescuers). The graph is composed of a *Generator* module that periodically produces sensor data (e.g. water surface elevation and water speed) for each point of a 2D discretization of the emergency scenario (e.g. a river basin). Each point is considered as an independent task by an adaptive *Solver* ParMod, that numerically solves a system of differential equations describing the flow behavior at the surface. Results evaluation is exploited by a *Client* component that performs post-processing and visualization activities on the mobile devices of the users.

An example of flood forecasting model is the TUFLOW hydrodynamic model Charteris et al. (2001), which is based on mass and momentum partial differential equations to describe the flow variation at the surface. Their discrete resolution requires, for each time slice, the resolution of a very large number of sparse tri-diagonal linear systems. For their resolution several highly scalable techniques have been developed over the last decades. In (Bertolli, Buono, Mencagli & Vanneschi, 2009) we have described two sequential variants of the *cyclic reduction* algorithm (Hockney & Jesshope, 1988).

We have provided two distinct parallel versions of the Solver ParMod: the first variant of the cyclic reduction, which requires a smaller number of operations than the second one, is executed inside a task-farm scheme, in which an emitter entity schedules input tasks (points) among a set of workers that execute in parallel the computation on different stream elements. The second version consists in a data-parallel version of the second variant. Although this algorithm requires a greater number of operations, data dependencies and also a higher memory utilization, it is amenable to be implemented in parallel through data-parallel techniques, in which a set of workers cooperatively compute the calculation on a partition of the same input data-structure. Moreover, in order to implement the data dependencies imposed by algorithm semantics, workers communicate with each other following a statically known communication stencil whose form varies between different iterations of the computation.

Such parallelizations respond to different QoS requirements: task-farm is able to increase the throughput of the computation in terms of number of completed points per time unit, whereas, the data-parallel scheme, can also improve the computation latency for completing a single task elaboration.

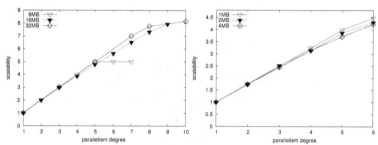

Fig. 4. Scalability of different versions of the flood forecasting model: task-farm on a cluster (left), data-parallel on a multicore architecture (right).

The two versions of the Solver ParMod are suitable for the execution on different computing architectures. The task-farm structure does not require intensive data exchanges between processes, therefore it can be efficiently implemented on a distributed-memory architecture (e.g. a cluster of production workstations). The data-parallel approach is instead amenable of being executed on a shared-memory architecture as a multicore platform. Scalability of the two versions is depicted in Figure 4, showing how different parallel versions achieve near optimal scalability results on highly different computing architectures.

Since the task-farm is based on replication of input data of different stream elements whereas the data-parallel applies a partitioning of the input data of the same task, we expect a different memory utilization of the two parallelization approaches. As described in (Bertolli et al., 2010), models describing the actual memory utilization of a structured parallel computation in function of its configuration parameters, as the parallelism degree and the parallelism scheme, provide accurate results as we can observe in the experiments depicted in Figure 5.

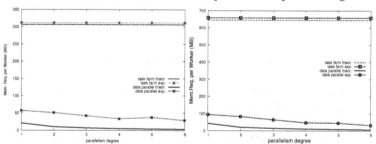

Fig. 5. Memory utilization per worker with tasks of size 16 MB (left) and 32 MB (right): comparison between task-farm and data-parallel schemes.

The experienced memory behavior follows the predicted values: i.e. for the task-farm scheme, based on a pure replication, the worker memory usage is independent on the parallelism degree; for the data-parallel scheme, based on a partitioning of input data, increasing the parallelism degree results on smaller partitions and thus on a lower memory usage per worker. This demonstrates how a proper versioning of a parallel module is useful not only for exploiting in the best way as possible different architectures on which the computation can be

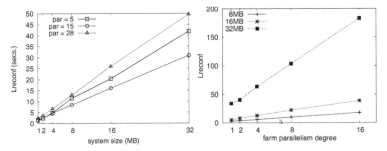

Fig. 6. L_{reconf} for the rollforward (left) and the rollback (right) switching form a source version (a task-farm) to a target version (a data-parallel) of the Solver ParMod. For the rollforward protocol L_{reconf} is dominated by the time spent for completing all the pending tasks in the source version, before starting the target one. For the rollback protocol L_{reconf} indicates the time necessary for re-executing all the rolled-back tasks.

executed, but also the memory utilization of a parallel program can be adapted in response to the current memory availability of the underlying execution platform.

In (Bertolli, Mencagli & Vanneschi, 2009) we have discussed adaptation strategies that modify the Solver behavior in response to the actual availability of network and computing resources. Especially for functional reconfigurations, in which the ParMod computation is migrated from a source to a target environment changing the parallelism version, we need to precisely estimate the overhead induced by these activities over the execution, in order to establish if a reconfiguration action is useful for the execution or not. In (Bertolli et al., 2011) we have presented different reconfiguration protocols for exploiting functional reconfigurations. A *rollforward protocol* consists in switching from the source to the target version only when all the pending tasks in execution in the source version of the ParMod have been completely processed. On the other hand a *rollback technique* minimizes the intervention of the source version on the protocol: i.e. when a reconfiguration is started, the target version starts its execution immediately, re-executing possible tasks that have been completed (or they are in execution) in the source version of the ParMod. Also in this case, based on the structured parallel programming paradigm, we are able to quantify the overhead induced by different protocols and choosing the best one for each specific case. Figure 6 depicts the time spent for applying a reconfiguration (i.e. *reconfiguration latency L_{reconf}*) in function of the task size.

Finally the exploitation of advanced control-theoretic techniques for controlling the Solver computation has been described in (Mencagli & Vanneschi, 2011) for a Cloud Computing scenario. We have analyzed the problem of executing the task-farm version of the Solver ParMod on a remote cloud architecture on which the cost of execution of a parallel computation is proportional to the number of virtualized resources (e.g. CPU) effectively used. Moreover, in order to discourage too many resource re-organizations, we have supposed the existence of a business model in which a fixed cost should be paid each time a new resource request is submitted to the cloud system.

In this scenario the design of effective adaptation strategies for optimizing the Solver execution are of great importance. For this application we have supposed a QoS objective with a twofold nature. First of all we need to complete the forecasting computation in the minimum

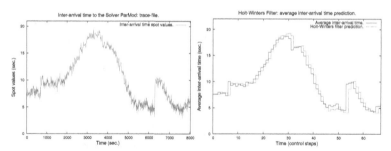

Fig. 7. Trace-file of the inter-arrival time experienced during the execution (left) and the statistical predictions obtained with an Holt-Winters filtering (right).

completion time as possible, such that emergency stakeholders can effectively plan proper response actions in advance to potentially dangerous events. Second we need to minimize the operational cost of the Solver execution on the cloud environment.

For this reason we have adopted the model-based predictive control strategy for the Solver ParMod, starting from a hybrid modeling of interesting QoS measurements of this computation (e.g. number of completed tasks and current operational cost). Due to a time-varying network availability, we have assumed a dynamic workload consisting in a variable mean inter-arrival time[2] of tasks from the Generator module. In Figure 7(left) is depicted the inter-arrival time of tasks experienced during an execution scenario in which the network that interconnects the Generator and the Solver ParMods alternates congestion phases and high-reliability phases (through the exploitation of an advanced network simulation tool, see (Mencagli & Vanneschi, 2011) for further details). In our modeling we have supposed the mean inter-arrival time as a disturbance input that needs to be predicted over a limited horizon. To do this we have exploited a *Holt-Winters filter* which is able to capture non-stationarity processes (e.g. trends and level shifts) of the underlying time-series of past inter-arrival time observations, achieving good prediction results as shown in Figure 7(right).

The task-farm scheme is a parallelization pattern featuring a clear and well-defined performance model: i.e. fixing the task-farm configuration in terms of mean inter-arrival time, service times of the emitter, the collector functionalities, and the number of parallel workers, we are able to analyze the task-farm performance behavior as a queueing network and evaluating the mean inter-departure time of results from this parallel computation.

This performance modeling can be exploited in order to predict how the number of computed tasks evolves during the execution. In (Mencagli & Vanneschi, 2011) the MPC approach has been applied providing an utility function that describes a proper trade-off between completed tasks and operational cost. Limited prediction horizons (e.g. 1, 2 and 3 control steps) have been applied for this experiment.

Figure 8(left) compares different adaptation strategies: (i) a reactive approach as the one described in (Bertolli, Buono, Mencagli & Vanneschi, 2009), in which the parallelism degree is modified according to the actual performance measurements of the computation; (ii) a

---

[2] For mean inter-arrival time we intend the average time between the reception of two subsequent input tasks from the Generator.

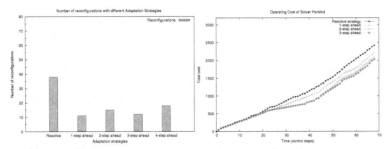

Fig. 8. Number of parallelism degree variations (left) and long-term operational cost of the Solver ParMod execution (right).

model-based predictive control technique that exploits the QoS predictability of the task-farm scheme. As we can observe the predictive approach has a positive impact on the *stability degree* of a ParMod configuration: i.e. for every prediction horizon length, the MPC strategy always features a lower number of reconfigurations than the reactive approach.

Figure 8(right) depicts the long-term operating cost throughout the execution. The importance of having a predictive run-time parallelism degree adaptation is clearly highlighted. W.r.t a purely reactive adaptation, the MPC strategy is able to reduce the operating cost of even the 20% if we exploit a prediction horizon of three steps, thus demonstrating how taking reconfigurations in advance to future workload predictions can be an effective adaptation technique.

7. Conclusion

In this chapter we have provided the design principles of a novel programming model approach for time-critical distributed applications (and notably emergency management systems). Such applications strongly require properties like the predictability of QoS parameters in function of the actual application configuration expressed in terms of identified software components, their interconnections, where they are currently deployed and how their computations are actually exploited. In order to meet these requirements our approach is based on a central point: the employment of the Structured Parallel Programming methodology.

Structured parallelism schemes are based on well-defined parallelization paradigms in terms of data replication or partitioning and function replication. We are able to define multiple alternative versions of the same software component performing critical tasks as forecasting models and simulations, as well as decision support systems. Such versions, besides being characterized by a predictable QoS behavior expressed through an analytical analysis (e.g. Queueing Network models), are also an effective way to deal with heterogeneous set of computing resources where the computation can be alternatively deployed. Moreover, this structured approach is amenable to provide predictable reconfiguration costs (e.g. from migrating the computation between different architectures) and also for being controlled through control-theoretic techniques (as the model-based predictive control procedure), in which the QoS predictability is exploited to plan reconfiguration actions in advance to critical events and optimizing the system execution and its operating cost.

8. References

Ahumada, S., Apvrille, L., Barros, T., Cansado, A., Madelaine, E. & Salageanu, E. (2007). Specifying fractal and gcm components with uml, *SCCC '07: Proceedings of the XXVI International Conference of the Chilean Society of Computer Science*, IEEE Computer Society, Washington, DC, USA, pp. 53–62.

Aldinucci, M., Campa, S., Danelutto, M. & Vanneschi, M. (2008). Behavioural skeletons in gcm: Autonomic management of grid components, *Parallel, Distributed and Network-Based Processing, 2008. PDP 2008. 16th Euromicro Conference on*, pp. 54–63.

Aldinucci, M., Coppola, M., Danelutto, M., Vanneschi, M. & Zoccolo, C. (2005). Assist as a research framework for high-performance grid programming environments, *Grid Computing: Software environments and Tools*, Springer, pp. 230–256.

Aldinucci, M., Danelutto, M. & Kilpatrick, P. (2009). Autonomic management of non-functional concerns in distributed & parallel application programming, *IPDPS '09: Proceedings of the 2009 IEEE International Symposium on Parallel&Distributed Processing*, IEEE Computer Society, Washington, DC, USA, pp. 1–12.

Aldinucci, M., Danelutto, M. & Vanneschi, M. (2006). Autonomic qos in assist grid-aware components, *PDP '06: Proceedings of the 14th Euromicro International Conference on Parallel, Distributed, and Network-Based Processing*, IEEE Computer Society, Washington, DC, USA, pp. 221–230.

Arshad, N., Heimbigner, D. & Wolf, A. L. (2007). Deployment and dynamic reconfiguration planning for distributed software systems, *Software Quality Control* 15(3): 265–281.

Berman, F., Fox, G. & Hey, A. J. G. (2003). *Grid Computing: Making the Global Infrastructure a Reality*, John Wiley & Sons, Inc., New York, NY, USA.

Bertolli, C. (2009). *Fault tolerance for High-Performance applications using structured parallelism models*, VDM Verlag, SaarbrÃijcken, Germany.

Bertolli, C., Buono, D., Mencagli, G. & Vanneschi, M. (2009). Expressing adaptivity and context-awareness in the assistant programming model, *Proceedings of the Third International ICST Conference on Autonomic Computing and Communication Systems*, pp. 38–54.

Bertolli, C., Mencagli, G. & Vanneschi, M. (2009). Adaptivity in risk and emergency management applications on pervasive grids, *Proceeding of the 10th International Symposium on Pervasive Systems, Algorithms, and Networks*, p. to appear.

Bertolli, C., Mencagli, G. & Vanneschi, M. (2010). Analyzing memory requirements for pervasive grid applications, *Parallel, Distributed, and Network-Based Processing, Euromicro Conference on* 0: 297–301.

Bertolli, C., Mencagli, G. & Vanneschi, M. (2011). Consistent reconfiguration protocols for adaptive high-performance applications, *Wireless Communications and Mobile Computing Conference (IWCMC), 2011 7th International*, pp. 2121–2126.

Bertsekas, D. P. (1995). *Dynamic Programming and Optimal Control, Two Volume Set*, 2nd edn, Athena Scientific.

Charteris, A., Syme, W. & Walden, W. (2001). Urban flood modelling and mapping 2d or not 2d, *Proceedings of the 6th Conference on Hydraulics in Civil Engineering: The State of Hydraulics*, Barton A.C.T: Institution of Engineers, Barton, Australia, pp. 355–363.

Cole, M. (2004). Bringing skeletons out of the closet: a pragmatic manifesto for skeletal parallel programming, *Parallel Comput.* 30(3): 389–406.

Coppola, M., Danelutto, M., Tonellotto, N., Vanneschi, M. & Zoccolo, C. (2007a). Execution support of high performance heterogeneous component-based applications on the grid, *Euro-Par'06: Proceedings of the CoreGRID 2006, UNICORE Summit 2006, Petascale Computational Biology and Bioinformatics conference on Parallel processing*, Springer-Verlag, Berlin, Heidelberg, pp. 171–185.

Coppola, M., Danelutto, M., Tonellotto, N., Vanneschi, M. & Zoccolo, C. (2007b). Execution support of high performance heterogeneous component-based applications on the grid, *Euro-Par'06: Proceedings of the CoreGRID 2006, UNICORE Summit 2006, Petascale Computational Biology and Bioinformatics conference on Parallel processing*, Springer-Verlag, Berlin, Heidelberg, pp. 171–185.

Danelutto, M., Vanneschi, M. & Zoccolo, C. (2007). A performance model for stream-based computations, *PDP '07: Proceedings of the 15th Euromicro International Conference on Parallel, Distributed and Network-Based Processing*, IEEE Computer Society, Washington, DC, USA, pp. 91–96.

der, S. A. J. v. & Schumacher, J. M. (1999). *Introduction to Hybrid Dynamical Systems*, Springer-Verlag, London, UK.

Garcia, C. E., Prett, D. M. & Morari, M. (1989). Model predictive control: theory and practice a survey, *Automatica* 25: 335–348.
URL: *http://portal.acm.org/citation.cfm?id=72068.72069*

Garlan, D., Siewiorek, D., Smailagic, A. & Steenkiste, P. (2002). Project aura: Toward distraction-free pervasive computing, *IEEE Pervasive Computing* 1(2): 22–31.

Gomes, A. T. A., Batista, T. V., Joolia, A. & Coulson, G. (2007). Architecting dynamic reconfiguration in dependable systems, pp. 237–261.

Hansmann, U., Merk, L., Nicklous, M. S. & Stober, T. (2003). *Pervasive Computing : The Mobile World (Springer Professional Computing)*, Springer.

Hellerstein, J. L., Diao, Y., Parekh, S. & Tilbury, D. M. (2004). *Feedback Control of Computing Systems*, John Wiley & Sons.

Hillman, J. & Warren, I. (2004). An open framework for dynamic reconfiguration, *ICSE '04: Proceedings of the 26th International Conference on Software Engineering*, IEEE Computer Society, Washington, DC, USA, pp. 594–603.

Hockney, R. W. & Jesshope, C. R. (1988). *Parallel Computers Two: Architecture, Programming and Algorithms*, IOP Publishing Ltd., Bristol, UK, UK.

Huebscher, M. C. & McCann, J. A. (2008). A survey of autonomic computing—degrees, models, and applications, *ACM Comput. Surv.* 40(3): 1–28.

Liberzon, D. (2003). *Switching in Systems and Control*, BirkhÃd'user Boston Production, Boston, USA.

Lillethun, D. J., Hilley, D., Horrigan, S. & Ramachandran, U. (2007). Mb++: An integrated architecture for pervasive computing and high-performance computing, *RTCSA '07: Proceedings of the 13th IEEE International Conference on Embedded and Real-Time Computing Systems and Applications*, IEEE Computer Society, Washington, DC, USA, pp. 241–248.

Lusk, E. (2002). Mpi in 2002: has it been ten years already?, p. 435.

Mathias, E., Baude, F. & Cave, V. (2008). A gcm-based runtime support for parallel grid applications, *CBHPC '08: Proceedings of the 2008 compFrame/HPC-GECO workshop on Component based high performance*, ACM, New York, NY, USA, pp. 1–10.

Mencagli, G. & Vanneschi, M. (2011). Qos-control of structured parallel computations: a predictive control approach, *Technical Report TR-11-14*, Department of Computer Science, University of Pisa, Largo B. Pontecorvo, 3, I-56127, Pisa, Italy.

Mortier, R. & Kiciman, E. (2006). Autonomic network management: some pragmatic considerations, *Proceedings of the 2006 SIGCOMM workshop on Internet network management*, INM '06, ACM, New York, NY, USA, pp. 89–93. URL: *http://doi.acm.org/10.1145/1162638.1162653*

Noble, B. D., Satyanarayanan, M., Narayanan, D., Tilton, J. E., Flinn, J. & Walker, K. R. (1997). Agile application-aware adaptation for mobility, *SIGOPS Oper. Syst. Rev.* 31(5): 276–287.

Russell, S. J. & Norvig, P. (2003). *Artificial Intelligence: A Modern Approach*, Pearson Education.

Snir, M., Otto, S. W., Walker, D. W., Dongarra, J. & Huss-Lederman, S. (1995). *MPI: The Complete Reference*, MIT Press, Cambridge, MA, USA.

Tsai, W. T., Song, W., Chen, Y. & Paul, R. (2007). Dynamic system reconfiguration via service composition for dependable computing, *Proceedings of the 12th Monterey conference on Reliable systems on unreliable networked platforms*, Springer-Verlag, Berlin, Heidelberg, pp. 203–224.

Vanneschi, M. (2002). The programming model of assist, an environment for parallel and distributed portable applications, *Parallel Comput.* 28(12): 1709–1732.

Vanneschi, M. & Veraldi, L. (2007). Dynamicity in distributed applications: issues, problems and the assist approach, *Parallel Comput.* 33(12): 822–845.

4

Emergency Evacuation Planning for Highly Populated Urban Zones: A Transit-Based Solution and Optimal Operational Strategies

Yue Liu[1] and Jie Yu[2]
[1]University of Wisconsin-Milwaukee/
[2]Shandong University
[1]USA
[2]China

1. Introduction

Natural and man-made disasters (e.g. hurricanes, floods, terrorist attacks) could cause huge economic loss and society damage. In many hazardous events, the best option is to relocate threatened populations to safer areas, which is commonly referred as emergency evacuation. During the process of evacuation, people would usually use their own vehicles to evacuate from the impacted area. However, there are some cases where people may not have access to reliable personal vehicles or using personal cars are not possible; then they need to rely on other forms of transportation. There are different modes of transportation that can be used to evacuate people, such as public transit, school buses, charter buses, demand-responsive vans, rail, and ambulances.

In view of literature, though significant contribution has been made in evacuation modelling considering passenger cars only, there are only a limited number of quantitative studies discussing the use of transit to evacuate the people during emergency management. One stream of researchers has employed simulation-based tools to study the feasibility and performance of transit evacuation plans. Liu et al. (2007, 2008) have presented an integrated system that embeds the evacuation of carless people; however the transit demand is converted into passenger car traffic in their system. Elmitiny et al. (2007) have performed a traffic simulation based study to evaluate alternative plans for the deployment of transit during an emergency situation in a transit facility such as a bus depot. Evacuation strategies evaluated include traffic diversion, bus signal optimization, access restriction, different destinations, and evacuation of pedestrians. Naghawi and Wolshon (2011a, 2011b) conducted a simulation-based assessment of the performance of the multi-modal evacuation traffic networks. The simulation results have shown that buses were able to increase the total number of people evacuated from the threat area while adding average queue length on some interstate freeway segments. Mastrogiannidou et al. (2009) developed an effective integration of the micro-simulation software package (VISTA) with transit based emergency evacuation models. A heuristic was developed to assign vehicle(s) to pickup points based on the shortest time criterion. They also study the impact of different numbers of available buses on routing strategies.

Another group of researchers have developed mathematical optimization models to obtain the best transit evacuation strategy. Perkins et al. (2001) discussed the use of buses to evacuate people (elderly and disabled) under a no-notice scenario. They assume that buses are at a garage, and optimize departure time of buses from the garage to pickup points such that the total travel time of buses is minimized. However, the routing strategy is static in their model and each bus would travel on a pre-set route to leave the affected area. No number of evacuees for each pickup point is mentioned. Sayyady (2007) has formulated the carless evacuation problem with a minimum cost flow model under additional side constraints. Their model assumes that bus stops are the pickup locations and the carless are guided to the stop that is closest to their current location waiting for pickup during an emergency. A Tabu search technique is used to identify evacuation routes for buses. In those studies, buses will only carry out one single trip and will not return to pick up the carless after leaving the affected area. Further studies (Tunc et al. 2011, Sayyady and Eksioglu, 2010) have also developed mixed-integer linear programming models to find optimal evacuation routes for transit.

Margulis et al. (2011) develop a binary integer-programming model to determine the assignment of buses to pickup points and to shelters during an evacuation. The objective of their model is to maximize the number of evacuee throughput in a given time period. However, their model assumes buses are at the pickup points at the beginning of the evacuation, and regulate each bus to return to the same evacuation site. He et al. (2009) has developed a stochastic optimization model to generate evacuation plans for transit-dependent residents in the event of a natural disaster. Their formulation features a location-routing problem (LRP) framework and solves for the number of shelters, their locations, the number of buses required, and their routes with the objective to minimize the total evacuation time. Comparative studies have also been performed to analyse single-stage and two-stage transit evacuation strategies. However, their assumption that all buses are at shelters might not be appropriate. Chen and Chou (2009) developed an optimal waiting spots and service locations selection model for transit-based emergency evacuation planning, and study the impact of transit-based evacuation to a highly dense populated area based on effectiveness measures such as network clearance time, move time, delay time, total travel time, and average speed of the traffic.

A very recent study by Chan (2010) has proposed a two-stage model for carless evacuation including a location problem that aims at congregating the carless at specific locations and a routing problem with the objective to pick up the carless from these evacuation sites and deliver them to safe locations. They explicitly consider the dynamic demand pattern of evacuees to pick up points as well as multiple trips of buses from pickup points to shelters. However, how to optimally guide evacuees to pick up points to better utilize the available buses is not discussed in their model.

Despite the significant contribution of previous studies in transit-based evacuation, none of those studies have integrated the dynamic processes of evacuee guidance (from buildings or parking lots to pick up points) and bus routing (from pick up points to shelters). Such integration will significantly improve the performance of the transit routing in response to the evacuee demand variation and maximize the utilization of available number buses by dynamically adjusting the demand distribution of evacuees at pick up points. In response to such critical research and operational needs, this study will propose an integrated

Emergency Evacuation Planning for Highly Populated Urban Zones: A Transit-Based
Solution and Optimal Operational Strategies

61

optimization model that is capable of coordinating the evacuee guidance and transit routing process seamlessly and simultaneously.

This chapter is organized as follows. The problem and the proposed mathematical model are presented in section 2. An algorithm to solve the given model is provided in section 3 of this chapter. Section 4 gives an illustrative example of the validity of mathematical model along with a brief sensitivity analysis on objective function weights. The chapter is finalized by concluding remarks in section 5.

2. Methodology

2.1 Problem description

As aforementioned, the proposed problem features a two-level optimization framework. The first level model guides evacuees from buildings and parking lots to designated pick-up points (e.g. bus stops, metro stations), and the level-II dispatches and routes buses from depots to pick-up points and transport evacuees to their destinations or safe places. This two-level problem can be converted into a graph form as shown in Figure 1. In this figure, nodes represent parking lots, pickup points, depots and safer area and arcs connecting those nodes represent the road network. The aim of the proposed of mathematical model is to find a sub-graph which is optimal with respect to maximizing the efficiency of the evacuation. In this network, evacuees are assigned to pickup points based on capacity of pickup points and distance, and once the demand is known the evacuation route for each bus will be constructed to transfer evacuees waiting at the pickup points to the safer area (shown in green circle in Figure 1). Considering the nature of this problem, we formulate it as a combined vehicle routing and assignment problem.

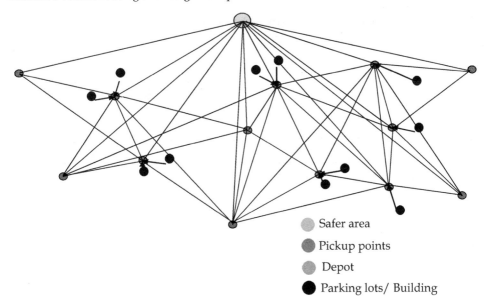

Fig. 1. Graphical representation of the two-level evacuation problem.

2.2 Assumptions

To ensure that the proposed formulations for the proposed problem are tractable and also to realistically reflect the real-world constraints, this study has employed the following assumptions in the model development.

- Evacuee walking time from buildings/parking lots to pickup points and bus travel time among pickup points;
- There exists a super evacuation destination in the network;
- The location and capacity of each pickup point is known;
- The capacity of buses are known as a priori; and
- Buses are restricted to go back to the same deport after sending evacuees to the destination.

2.3 Notation

To facilitate model presentation, notations used hereafter are summarized in Table 1.

I	Set of parking lots/buildings
L	Set of depots
H	Set of pickup points
H_0	Set of pickup point and the evacuation destination
K	Set of buses
d_{ip}	Distance between parking lot/building i and pickup point p
l_{pq}	Distance between pickup point p and q
S_{ped}	Evacuee walking speed
Di	Evacuation demand at parking lot/building i
C_p	Capacity of pickup point p, $p \in H$
Q_k	Capacity of bus k
w_i	Weight of expression i in the objective function
X_{ip}	Number of passengers/evacuees at building i assigned to pickup points p
Y_{pqk}	1- If arc (p, q) belongs to the route operated by bus k 0- Otherwise
T_{pqk}	Number of evacuees at pickup point q assigned to bus k goes from p to q
U_{pk}	An auxiliary variable for sub-tour elimination constraint in route k

Table 1. Notation of parameters and variables for the mathematical model

2.4 Model formulation

The two-level evacuation problem can be formulated as the following mixed integer program (MIP):

$$\text{Minimize} \quad w_1 \sum_p \sum_q \sum_k l_{pq} Y_{pqk} - w_2 \sum_i \sum_p \frac{S_{ped}}{d_{ip}} X_{ip} \qquad (1)$$

s.t.

Emergency Evacuation Planning for Highly Populated Urban Zones: A Transit-Based
Solution and Optimal Operational Strategies

63

$$\sum_{k \in K} \sum_{p \in L \cup H_0} Y_{pqk} \geq 1 \quad \forall q \in H_0 \tag{2}$$

$$U_{pk} - U_{qk} + |H| \times Y_{pqk} \leq |H| - 1 \quad p,q \in H_0, k \in K \tag{3}$$

$$\sum_{p \in L \cup H} Y_{pqk} - \sum_{p \in L \cup H_0} Y_{qpk} = 0 \quad q \in L \cup H_0, k \in K \tag{4}$$

$$\sum_{p} \sum_{q \in L \cup H_0} Y_{pqk} \leq 1 \quad \forall k \in K \tag{5}$$

$$\sum_{k \in K} \sum_{p \in L \cup H_0} Y_{pqk} \geq 1 \quad \forall q \in L \tag{6}$$

$$\sum_{p \in H} X_{ip} = D_i \quad \forall i \in I \tag{7}$$

$$\sum_{i \in I} X_{ip} \leq C_p \quad \forall p \in H \tag{8}$$

$$\sum_{p} \sum_{q} T_{pqk} \leq Q_k \quad p,q \in H_0 \quad \forall k \in K \tag{9}$$

$$T_{pqk} \leq Q_k Y_{pqk} \quad \forall p \in H \cup L, \forall q \in P, \forall k \in K \tag{10}$$

$$\sum_{p \in L \cup H} \sum_{k \in K} T_{pqk} - \sum_{i \in L} X_{iq} \geq 0 \quad \forall q \in H \tag{11}$$

In this formulation the objective function is defined as Eq. (1). The objective function includes two terms: the first term deals with routing and the second one is related to assigning evacuees to the pickup point. The first term minimizes the total distance and second term tries to maximize total number of evacuees assigned to pickup point considering the walking distance. Since the objective function finds the minimum possible value for total bus traveling distances and the maximum total number of evacuees assigned, it implicitly maximizes total evacuees transported over the shortest path for each bus, and therefore maximizes the number of evacuees to the safer area.

The number of buses dispatched from each depot must be at least one which as given by constraint (2); Constraint (3) is used for sub-tour elimination in the VRP problem; Constraint (4) ensures flow conservation of network; Constraint (5) guarantees that each bus can be utilized at most once during one round of the evacuation period; Constraint (6), unlike conventional constraints in the VRPs, ensures that each link can be served by more than one bus, even for those leaving from pickup points.

Eq. (7) is the first assignment-type constraint of the model, which guarantees all evacuees must be assigned to pickup points. Moreover, this number must be less than each pickup point capacity constrained by Eq. (8). Constraint (9) limits the number of evacuees transferred from pickup points to the evacuation destination must be less than the bus capacity during each tour or route.

Constrain (10) relates assignment of evacuees to buses only if the bus serves that link or pickup points. Constraint (11) guarantees that all evacuees at pickup points are assigned to vehicles and transferred to the evacuation destination.

3. Solution algorithm

Note that the proposed formulation of the model is a NP-hard problem, solution of the large-scale instances are intractable. To ensure the applicability of the proposed model in real-world scenarios, this section develops a two-stage Tabu-search-based approach to solve the model.

In the first stage, a relaxed assignment problem is solved to find the evacuee demand at each pickup point based on which a route for each bus is constructed through a meta-heuristic algorithm. The second stage is a Tabu search meta-heuristic which solves the VRP sub-problem. The flowchart for the proposed solution algorithm is depicted in Figure 2 with each of its elements explained as follows:

3.1 Parameter initialization

Before implementing the heuristic, some parameters should be known and initialized in advance. These parameters are the number of buildings, pickup points, depots, buses, capacity of each, and size of TABU list.

3.2 Stage-1: Solve the assignment problem with relaxation

At this step, the assignment part of the problem is considered. In order words, those constraints related to the routing part of the problem are relaxed; then a solver is used to solve the assignment problem. The output of this step is the number of evacuees that are assigned to the pickup points waiting for buses.

3.3 Stage-2: Tabu search

3.3.1 Step 1- initial solution generation

To generate the initial solution, we have developed the following steps:

- Step 1.1: A distance-based proximity matrix for each pair of nodes is developed;
- Step 1.2: Find the minimum value between depots and pickup points;
- Step 1.3: Based on the waiting evacuees and bus capacity, assign the maximum possible evacuees to the bus; go to step 1.4;
- Step 1.4: If bus has room for more evacuees, go to step 1.5; otherwise go to step 1.6;
- Step 1.5: Add the nearest pickup point to the route and go to step 1.4;
- Step 1.6: The bus goes to the safer area to drop off passenger; go to step 1.8;
- Step 1.7: If there are evacuees waiting at any pickup point, go to step 1.2; otherwise go to step 1.6;
- Step 1.8: END.

3.3.2 Step 2- neighborhood generation

In order to search the solution space, some solution, known as a neighbour, must be generated from the current solution. If the newly generated solution is better than current

Emergency Evacuation Planning for Highly Populated Urban Zones: A Transit-Based
Solution and Optimal Operational Strategies

65

one, a move will be made and that move will be added to a list called the Tabu list, which is necessary to avoid from falling back to same point and local optima. By definition, a solution S' that does not include TABU moves is the neighbour of solution S if it is feasible to the problem and represents an adjacent flow. The TABU search algorithm implements the 2-opt search mechanism to find a better solution (Cordeau et al., 1997).

In the proposed problem, a neighbour is defined based on the order of the pickup points served by a bus or exchanging two pickup points in two different routes. However, this may result in an infeasible solution due to the bus capacity violation constraint. Therefore, to make sure the neighbour solution is feasible; one needs to check the capacities of two buses before making the move, and then swap the two pickup points. Assume, without loss of generality, vehicles are v_1 and v_2 and pickup point in v_1 route is p_1 with number of evacuees assigned T_1 and pickup point in v_2 route is p_2 with number of evacuees assigned T_2. If $T_1 > T_2$, two pickup points are swapped and another p_1 is added to v_1 to pick up the remaining evacuees at p_1. The same procedure will apply to the other case, which will keep the heuristic from generating infeasible solutions.

3.3.3 Step 3- selection strategy

The selection strategy determines the rule for selecting the next neighbouring solution. In this study, we employ the first better move strategy (Cordeau et al., 1997), in which the neighbouring solutions are investigated in a predetermined order, and the first solution that shows an improvement is selected as the next solution.

3.3.4 Step 4- TABU list

Since in this heuristic, contrary to classical methods, the current solution may deteriorate from one iteration to the next iteration, recycling may occur. To avoid this some recently explored solutions, bus routes in our case, are temporarily declared as Tabu or forbidden. We use the TABU list (θ) defined as a finite list with fixed size containing TABU sub-paths. When a move is made based on the selection strategy, that move is added to the list to make sure that the algorithm does not go to the current solution. Since the size of the TABU-list is bound by L, when the $|\theta| = L$, the new one is added after removing the oldest one.

3.3.5 Step 5- evaluation criterion

Based on the first better strategy (Cordeau et al., 1997), the first neighbouring solution that improves the evaluation criterion is selected as the better solution. We simply define the evaluation criterion to be value of objective function for mathematical model to minimize.

3.3.6 Step 6- termination criterion

The TABU search algorithm implements the 2-opt mechanisms for a fixed number of iterations. If no better neighbouring solution is found, then it moves to the best neighbouring solution, even if it does not improve the current solution. Such moves are known as bad moves, *badMoves*. After making a fixed number of bad moves (*maxBad*), the algorithm stops by reporting the best-found solution over all iterations.

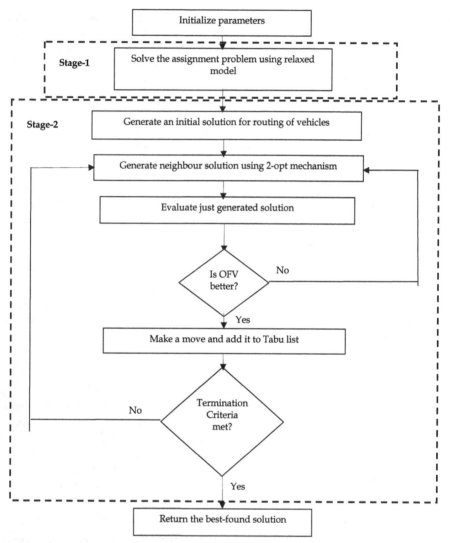

Fig. 2. Flowchart of the proposed heuristic algorithm.

In the next section, we will use a small-scale numerical example to validate the applicability and feasibility of the proposed mathematical model.

4. A numerical study

4.1 Test case description

In order to validate the structure and applicability of the proposed formulation, a numerical example is solved and discussed in this section. The data used in this example is given in Table 2. In all Tables 2(a)-(e), buildings, pickup points and depots are shown by initials.

Emergency Evacuation Planning for Highly Populated Urban Zones: A Transit-Based
Solution and Optimal Operational Strategies

67

(a) Number of nodes and vehicles of the numerical example

Parking Lots	Pickup Points	Depots	Vehicle	Vehicle capacity
10	6	3	8	45

(b) Distances from buildings (B) to pickup points (P) (unit: in 0.1 miles)

	P1	P2	P3	P4	P5	P6
B1	3	4	1	2	1	3
B2	2	1	3	1	1	1
B3	1	1	1	1	2	2
B4	1	2	2	1	2	3
B5	1	2	3	3	4	1
B6	3	4	1	2	1	3
B7	2	1	3	1	1	1
B8	1	1	1	1	2	2
B9	1	2	2	1	2	3
B10	1	2	3	3	4	1

(c) Distance matrix for vehicular network (unit: in 0.1 miles)

	P1	P2	P3	P4	P5	P6	D1	D2	D3
P1	1000	20	30	20	10	30	20	10	15
P2	20	1000	40	30	10	20	30	40	15
P3	30	40	1000	10	20	5	15	25	35
P4	10	20	30	1000	40	25	35	40	45
P5	20	10	40	30	1000	25	35	45	5
P6	20	20	40	30	10	1000	30	40	15
D1	30	40	15	10	20	5	1000	25	35
D2	20	40	40	30	10	20	30	1000	15
D3	10	20	30	45	40	25	35	40	1000

(d) Number of evacuees at each building (unit: # of evacuees)

Building/ Parking lot	B1	B2	B3	B4	B5	B6	B7	B8	B9	B10
Demand	10	40	36	64	14	10	40	36	64	14

(e) Capacity of each pickup point (unit: # of evacuees)

Pickup points	P1	P2	P3	P4	P5	P6
Capacity	80	60	70	60	50	80

Table 2. Data used in the numerical example.

Table 2(a) lists all problem indices, which will be used to solve the example. Entries in this table are self-explanatory. Table 2(b) depicts distances from buildings/parking lots to each pickup point. Distances between nodes of vehicular network (pickup points, depots and safer area) are given in Table 2(c), in which P's and D's stand for pickup points and depots. Tables 2(d) and 2(e) give the number of evacuees at each building/parking lot and the capacity of each pickup point, respectively. For example, the number of evacuees waiting at the first building is 10 and the first pickup point can accommodate no more than 80 evacuees.

Note that data used in the numerical example is for the purpose of validation of the proposed model and may not be realistic considering a real-world evacuation. However, the proposed model is generic and can handle real-world evacuation scenarios when the data is available.

4.2 Results and discussion

The numerical example is solved with CPLEX 11.2 in 704 seconds of computer time. The assignment of evacuees from buildings or parking lots to pickup points (the first level problem) and the bus routing plans among pickup points and depots (the second level problem) are solved concurrently with the proposed formulation. Eight buses are used to take evacuees to the safer area. In should be noted that since one term of objective function is related to route cost, the model indirectly minimized the number of buses used to evacuate carless people. A graphical illustration of the numerical example results is shown in Figure 3. In the figure, blue points are building/parking lot from which evacuees are assigned to pickup points (shown by red arrow in the figure). The bus routing plans that take evacuees from pickup points to the safer area and then come back to their depot are also illustrated in Figure 3. For instance, one route (bus) starts from depot 2 to pickup point 1, goes to safer area (shelter) and finishes its journey by coming back to its origin.

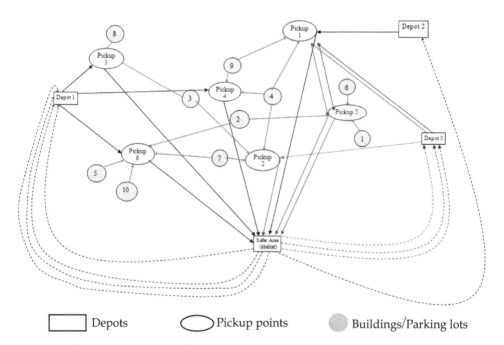

☐ Depots   ⬭ Pickup points   ⬤ Buildings/Parking lots

Fig. 3. Graphical representation of the numerical example results.

For the first level, evacuees are assigned based on the accessibility and available capacity at designated pickup points, as shown in Table 3. For instance, there are 36 evacuees waiting at building 3, 9 of them are assigned to pickup point 2 and the remaining 27 are assigned to pickup point 3. On the other hand, the capacity of pickup point 2 is 60, which take 9 evacuees from building 3, 3 from building 4, and other 33 from building 7.

	Pickup 1	Pickup 2	Pickup 3	Pickup 4	Pickup 5	Pickup 6	Total
Building 1					10		10
Building 2					30	10	40
Building 3		9	27				36
Building 4	30	3		31			67
Building 5						14	14
Building 6					10		10
Building 7		33				7	40
Building 8			36				36
Building 9	50			14			64
Building 10						14	14
Total	80	45	63	45	50	45	

Table 3. Assignment of evacuees from buildings/parking lots to pickup points (unit: # of evacuees).

For the second level, the routing plan for each bus during the evacuation process is summarized in Table 4. Also reported in Table 4 is the number of evacuees taken at each pickup point and transported to the evacuation destination by each bus. It can be observed that more than one bus has been assigned to each route depending on the number of evacuees. For example buses 4 and 5 in this table have the same route because number of evacuees at pickup point 1 is 80 and bus 8 will take 45, and the remaining 35 evacuees are transported by buses 4 and 5. This is due to the fact that the proposed problem structure allows multiple buses on each route, which is different from the assumption of traditional vehicle routing problem. Another notable fact is that the capacity of each bus is fully utilized. Since bus capacity is 45, if we look at results in Table 4, it can be observed that buses 2, 3, 6, 7 and 8 are used with full capacity and other buses carry less than capacity because the numbers of evacuees waiting in those places are less than bus capacity. For instance, bus 1 takes only 18 remaining evacuees at pickup point 3.

Buses	Routing Plan	#of Evacuees/Throughput
1	Depot 1 – Pickup 3 – Depot 1	18
2	Depot 1 – Pickup 4 – Depot 1	45
3	Depot 1 – Pickup 6 – Depot 1	45
4	Depot 3 – Pickup 1 – Pickup 5 – Depot 3	22 + 22 = 44
5	Depot 3 – Pickup 1 – Pickup 5 – Depot 3	13 + 28 = 41
6	Depot 1 – Pickup 3 - Depot 1	45
7	Depot 3 – Pickup 2 – Depot 3	45
8	Depot 2 – Pickup 1 – Depot 2	45

Table 4. The routing plan of each bus and the total of number of evacuees transported.

Based on the results given in Tables 3 and 4, it is apparently clear that the proposed mathematical model can solve this evacuation example to optimality, and both the evacuee assignment and bus routing plans generated from the model are valid. The next step is to check the power of the model in dealing with large and real size problems and to show how model parameters can affect outputs and the solutions. In addition, the validation of the model will give us some guidelines to design a heuristic algorithm to find good solution faster, which will be discussed in the next section in more details.

4.3 Sensitivity analysis on objective function weights assignment

For sensitivity analysis, the effect of different weights of objective function terms on the result and computational time for the above case is studied. Each term of objective function is assigned a weight in range [0, 1] summing up to one. The result of this test is tabulated in Table 5. It can be observed that the value of each term of objective function is not sensitive to its weight assignment, which indicates that the problem is not sensitive to the objective weights. On the other hand, there is one case in which total number of evacuees taken to safer area is different from others. While other cases have 328 evacuees as throughput, this case, in which $w_1 = 0.4$ and $w_2 = 0.6$, has 340 evacuees as throughput. This reveals the fact that there might be multiple solutions with the same objective value, if weights are not considered.

Weights		Value of objective term		Structure of the solution	Throughput (#of evacuees)	Time (sec.)
w_1	w_2	1	2			
1	0	305	0	V1: 25 V2: 45 V3: 45 V4: 45 V5: 35 V6: 45 V7: 45 V8: 35 + 8 =43	328	309
0.9	0.1	305	328	V1: 25 V2: 45 V3: 45 V4: 30 + 15 = 45 V5: 45 V6: 45 V7: 12 + 33 = 45 V8: 33	328	567
0.8	0.2	305	328	V1: 25 V2: 45 V3: 35 + 10 = 45 V4: 45 V5: 45 V6: 45 V7: 33 V8: 43 + 2 + 45	328	856

Emergency Evacuation Planning for Highly Populated Urban Zones: A Transit-Based
Solution and Optimal Operational Strategies

71

Weights		Value of objective term		Structure of the solution	Throughput (#of evacuees)	Time (sec.)
w_1	w_2	1	2			
0.7	0.3	305	328	V1: 35 V2: 45 V3: 30 V4: 45 V5: 45 V6: 45 V7: 45 V8: 25	328	557
0.6	0.4	305	328	V1: 43 V2: 35 V3: 25 V4: 28 + 17 = 45 V5: 45 V6: 45 V7: 45 V8: 45	328	392
0.5	0.5	305	328	V1: 45 V2: 30 + 15 = 45 V3: 45 V4: 17 + 28 = 45 V5: 45 V6: 35 V7: 45 V8: 23	328	352
0.4	0.6	305	328	V1: 25 V2: 45 V3: 45 V4: 45 V5: 15 + 30 =45 V6: 45 V7: 45 V8: 30 + 15 = 45	340	794
0.3	0.7	305	328	V1: 45 V2: 45 V3: 25 V4: 35 V5: 35 + 10 = 45 V6: 45 V7: 45 V8: 43	328	1051
0.2	0.8	305	328	V1: 45 V2: 25 V3: 27 + 18 = 45 V4: 45 V5: 43 V6: 45 V7: 35 V8: 45	328	734

Weights		Value of objective term		Structure of the solution	Throughput (#of evacuees)	Time (sec.)
w_1	w_2	1	2			
0.1	0.9	305	328	V1: 45 V2: 25 V3: 22+23=45 V4: 45 V5: 45 V6: 33 V7: 30+15=45 V8: 45	328	525

Table 5. Effect of objective weights on throughput.

The effect of weights on throughput (number of evacuees) for both terms in objective function is graphed in Figure 4. In this figure, the blue line is for the first term in the objective function, and the red one is for the second term in the objective function. As shown, there is one peak for each term. This peak happens when the weight for the first term (w_1) is 0.4 and weight for the second term (w_2) is 0.6 ($w_1 + w_2 = 1$), which yields the highest throughput (340 evacuees). During the process of evacuation, operators can use the proposed model and select proper sets of weights for the objective terms to achieve the expected evacuation system performance.

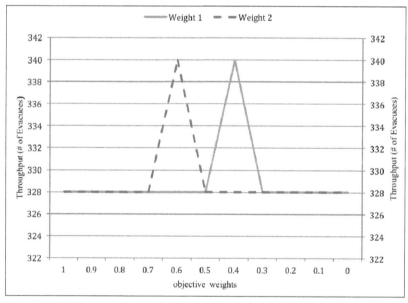

Fig. 4. Effect of objective terms weights on throughput (# of evacuees).

5. Conclusions

This chapter presents a mathematical model for evacuation planning in highly populated urban zones where a potentially large number of pedestrians depend on transit for

Emergency Evacuation Planning for Highly Populated Urban Zones: A Transit-Based
Solution and Optimal Operational Strategies

73

evacuation. The uniqueness of the proposed model lies in its capability to concurrently operate the dynamic processes of evacuee guidance (from buildings or parking lots to pick up points) and bus routing (from pick up points to shelters). Such integration will significantly improve the performance of the transit routing in response to the evacuee demand variation and maximize the utilization of available number of buses by dynamically adjusting the demand distribution of evacuees at pick up points.

The model is formulated as a combined vehicle routing and assignment problem and solved by a two-stage Tabu-based heuristic to yield meta-optimal solutions. The feasibility and applicability of the proposed model is illustrated with a numerical example solved to optimality. Results show that the proposed model can yield valid and detailed evacuee guiding and transit routing plans during the evacuation within a reasonable time window. Sensitivity analysis of the impact of objective function weights indicates that the proposed model is robust and not sensitive to the weight variations. It also provides guidelines for evacuation operators on best customizing the objectives to achieve expected evocation operational performance.

Note that the proposed model is only validated with a numerical test, and the results remain preliminary. Next step research will be testing the model's applicability in real-world evacuation scenarios. Computational performance of the proposed solution algorithm will also be evaluated. In addition, the problem studied here is static, in the way that a stable table of evacuee demand and number of buses during an evacuation period is given. The assignment of evacuees and routing of buses also use a static representation of the network condition. Extending the model to an explicitly dynamic setting, with time-varying demand generation rates and travel times, is another worthwhile direction for further work. From a computational standpoint, such extension further complicates an already complex problem, but the dynamic environment during evacuation requests this to be included.

6. References

Chan, C. P. (2010). *Large Scale Evacuation of Carless People During Short- and Long-Notice Emergency*, PhD Dissertation, Department of Systems and Industrial Engineering, The University of Arizona

Chen, C. & Chou, C. (January 2009). Modeling and Performance Assessment of a Transit-Based Evacuation Plan within a Contra-Flow Simulation Environment, *Proceedings of 88th Annual Meeting of the Transportation Research Board*, Paper Number: 09 – 2739, January 11 – 15, 2009

Cordeau; J-F; Gendreau, M. & Laporte, G. (1997). A Tabu Search Heuristic for Periodic and Multi-depot Vehicle Routing Problems, *Networks*, Vol.30, No.2, (September 1997), pp. 105-119, ISSN 0028-3045

Eksioglu, B.; Vural, A.V.; & Reisman, A. (2009). The Vehicle Routing Problem: A Taxonomic Review, *Computers & Industrial Engineering*, Vol. 57, (November 2009), pp. 1472-1483, ISSN: 0360-8352

Elmitiny, N.; Ramasamy, S. & Radwan, E. (2007). Emergency Evacuation Planning and Preparedness of Transit Facilities: Traffic Simulation Modeling, *Transportation Research Record*, No. 1992, (September 2007), pp. 121-126, ISSN 0361-1981

He, S.; Zhang, L.; Song, R.; Wen, Y. & Wu, D. (January 2009). Optimal Transit Routing Problem for Emergency Evacuations, *Proceedings of 88th Annual Meeting of the*

Transportation Research Board, Paper Number: 09 – 0931, Washington, D. C., January 11 – 15, 2009

Liu, Y.; & Chang, G. L. (October 2007). CAPEVACUATION: The Corridor-Based Emergency Traffic Evacuation System for Washington D.C., *14th World Congress on Intelligent Transportation Systems*, Beijing, China, 2007

Liu, Y. ; Chang, G. L. ; Liu, Y. ; & Lai, X. (2008). A Corridor-Based Emergency Evacuation System for Washington, D.C.: System Development and Case Study, *Transportation Research Record*, No. 2041, (August 2008), pp. 58-67, ISSN 0361-1981

Margulis, L.; Charosky, P. & Fernandez, J. (June 2011). Hurricane Evacuation Decision-Support Model For Bus Dispatch, 17.06.2011, Available from http://ormstomorrow.informs.org/archive/Summerfall06/

Mastrogiannidou, C.; Boile, M.; Golias, M.; Theofanis, S. & Ziliaskopoulos, A. (2009). Using Transit to Evacuate Facilities in Urban Areas: A Micro-Simulation Based Integrated Tool, *Proceedings of 88th Annual Meeting of the Transportation Research Board*, Paper Number: 09 – 3439, Washington, D. C., January 11 – 15, 2009

Naghawi, H. & Wolshon, B. (2011). Performance of Multi-Modal Evacuation Traffic Networks: A Simulation Based Assessment, *Proceedings of 90th Annual Meeting of the Transportation Research Board*, Paper Number: 11 – 1803, Washington, D. C., January 23 –27, 2011

Naghawi, H. & Wolshon, B. (2011). Operation of Multimodal Transport Systems During Regional Mass Evacuations, *Proceedings of 90th Annual Meeting of the Transportation Research Board*, Paper Number: 11 – 1843, Washington, D. C., January 23 –27, 2011

Perkins, J. A.; Dabipi, I. K.; & Han, L. D. (2001). Modeling Transit Issues Unique to Hurricane Evacuations: North Carolina's Small Urban and Rural Areas, Transportation Institute, North Carolina Agricultural and Technical State University, Urban Transit Institute, Transportation Institute

Sayyady, F. (2007). *Optimizing the Use of Public Transit System in No-Notice Evacuations in Urban Areas*, Master Thesis, Department of Industrial and Systems Engineering, Mississippi State University

Sayyady, F.; & Eksioglu, S.D. (2010). Optimizing The Use of Public Transit System During No-Notice Evacuation, *Computers & Industrial Engineering*, Vol. 59, No. 4, (November 2010), pp. 488-495, ISSN: 0360-8352

Tunc, H.; Eksioglu, B.; & Eksioglu, S. D. (May 2011). Optimizing the Use of Transit System with Information Updates during No-Notice Evacuations, *Proceedings of the Industrial Engineering Research Conference*, Reno, NV, 2011

5

Supply Allocation and Vehicle Routing Problem with Multiple Depots in Large-Scale Emergencies[*]

Jianming Zhu
College of Engineering, Graduate University of Chinese Academy of Sciences
China

1. Introduction

Large-scale emergencies, such as substantial acts of nature, large human-caused accidents, and major terrorist attacks, are of high-consequence, low-probability (HCLP) events that may result in loss of life and severe property damage. In recent years, developing decision-oriented operations research models to improve preparation for and response to major emergencies has drawn more and more attention (see Altay & Green (2006) and Larson et al. (2006)).

Relief resources play an important role in emergency management after disasters, such as medicine, food, tent, etc. Due to scarce resources and overwhelming demands during an emergency (especially in the early stages) careful pre-planning and efficient execution can save lives. A key factor in an effective response to an emergency is the prompt availability of necessary supplies at emergency sites. Therefore, efficient emergency logistics becomes important in addressing and optimizing the complex distribution process. In most real-life situations, the distribution process is typically divided into two decision stages. In the first stage, supply quantity allocated to each demand location is determined. This is referred to as the allocation problem. In the second stage, how supplies will be transported is determined, which may be modeled as a Vehicle Routing Problem. Obviously, when supply is large enough, the allocation problem is trivial, while the second stage is still a complex problem.

Traditional Vehicle Routing Problem (VRP) is to design the least cost routes for a vehicle fleet to supply goods from inventory to customer locations. The problem was first introduced by Dantzig & Ramser (1959) to solve a real-world application concerning the delivery of gasoline to service stations. A comprehensive overview of the VRP can be found in Toth & Vigo (2002) and other general surveys on the deterministic VRP can also be found in Laporte (1992). Various specific VRP models, e.g. with time windows, multiple depots, dynamic routes, and stochastic customer demands, etc. were published in Rathi et al. (1993) and Renaud et al. (1996). Astrid & David (2003) considered vehicle routing problem with random travel time and service time, while all vehicles departed from the same depot. Enrico & Maria (2002) considered periodic vehicle routing problem (PVRP) while vehicles can renew their capacity at some intermediate facilities. Recently, an exact algorithm was presented for PVRP in

[*]This work is partially supported by National Natural Science Foundation of China (71001099, 90924008) and by the President Fund of GUCAS.

Roberto et al. (2011). Dynamic request occurrence is considered in Lorini et al. (2011). Almost all VRP models and algorithms are for "normal operation" that minimize cost represented by travel distances or travel times and applied in daily operating logistic systems. Only several works considered total arrival time (see Campbell et al. (2008) and Ngueveu et al. (2010)).

The highly unpredictable nature of large-scale emergencies, unfortunately, leads to significant uncertainty both in demand and travel times. For example, in certain emergency cases, medication or antidotes must be applied within a specific time limit from the occurrence of the event to maximize their effectiveness to save lives. Requirement for the medication may change rapidly along with the case development and is hard to predict. Traditional pharmaceutical supply chains are no longer adequate to provide the rush demand. In emergency cases the so called Strategic National Stockpile, a large managed inventory from manufacturers, may be used. Vehicle fleet size can be uncertain due to emergency calls. The vehicles may load supply from multiple depots (e.g. airports) and may not return to the original depot location. Travel times of transporting the medication from the central supply to the demand population areas also become uncertain in case of emergency because of sudden road congestion and panic, or because of strict traffic control. Thus, the objectives of the VRP for response to emergency are usually to minimize both the unmet demand and delay time. Finally, an efficient algorithm to find a good solution is very important for emergency operation managers.

As discussed above, transportation is an important issue in emergency response, which is called emergency logistics. Emergency logistics management has also emerged as a worldwide-noticeable theme. Sheu (2007) presented four main challenges under which emergency logistics management can be characterized. Also as a sponsor, Sheu edited a special issue of Transportation Research Part E, in which six papers on emergency logistics were included. These papers concentrated on addressing the issue of relief distribution to affected areas. Consignment of supply is typically examined in the literature as a multi-commodity network flow problem, with a multi-period and/or multi-modal setting. Haghani & Oh (1996) formulated a multi-commodity, multi-modal network flow model with time windows for disaster response. Two heuristic algorithms were proposed. The flow of supply over an urban transportation network was modeled as a multi-commodity, multi-modal network flow problem by Barbarosoglu & Arda (2004). A two-stage stochastic programming framework is formed as the solution approach. Another study on the topic, conducted by Fiedrich et al.(2000), model the problem similar to a machine scheduling problem proposing two heuristics, Simulated Annealing and Tabu Search. Yi & Ozdamar (2004) considered a dynamic and fuzzy logistics coordination model for conducting disaster response activities. The model was illustrated on an earthquake data set from Istanbul. Also, Barbarosoglu & Arda (2004) proposed that their model could be used effectively within a decision-aid tool by public and non-public response agencies that are obscured by the variability of impact estimations under large number of different earthquake scenarios.

All these uncertain factors must be considered by an emergency operation manager in dispatching vehicles to effectively deliver the life-saving demands to the people in need. Due to the characteristics of uncertainty of large-scale emergency, a dynamic VRP can be stated as follows:

1. when an emergency occurs, with reported demand calls, a responder must evaluate the demand pattern, including locations, quantity, and time requirement for the deliveries.
2. organize the supplies and route the available vehicles to meet the emergency requirements in an efficient way to minimize the unmet demand and the total time delay.
3. with the updated demand information, relocate medical supplies and vehicles, route and dispatch next fleet with the same objective.
4. keep evaluating the updated demand and routing further vehicles, until all the demand is met.

Shen et al. (2007) studied a stochastic VRP model with time windows that minimize unmet demand for large-scale emergencies. In their paper, vehicle time delay is not allowed when visiting a demand node. However this strict limitation may be unreasonable because in emergency situations even the urgency of need for medical supplies may not be met from a time perspective. The dispatcher will still send the supply to save as many lives as possible with the least time delay. Liu et al. (2007) considered both the unmet demand and time delays.

This paper focuses on modeling and solution framework for the VRP in response to a large-scale emergency. In Section 2, a deterministic VRP model with multiple depots will be presented. In section 3, this model will be analyzed in detail. Then, an efficient heuristic algorithm is designed for the proposed model in Section 4. Finally in Section 5,, numerical experiments and a case simulation demonstrate that the model and algorithm can be very useful as a decision tool for emergency responders.

2. A deterministic VRP model with multiple depots

In this paper, we consider a situation that several fleets of vehicles send emergency supply from multiple depots (e.g. airport or central inventory) to demand locations (e.g. hospitals or triage stations), and return to the original depots after delivery all the supply. Objectives of the model is to minimize the maximum unsatisfied rate and the total weighted time delay.

According to emergency conditions, we have several assumptions:

1. There is limited amount of supply in each depot.
2. Each demand node has a deadline for supply, and delay is permitted.
3. The traveling time between each pair of nodes is deterministic.
4. Vehicles are not reusable.

In the model presented in this paper, the total weighted time delays are explicitly expressed in the objective function as the most important factor.

Now, decision variables and parameters will be specified.

Set D represents demand nodes. L is denoted as supply set including depots and other suppliers. We consider fleet sets $K(l)$ of vehicles at supplier l. Let $K = \bigcup_{l \in L} K(l)$ for simplification. The node set is expressed as $C = D \bigcup L$. Suppose from each node i to any other node j there is a route, or an arc (i, j). Therefore a transport network can be expressed by the node set $C = D \bigcup L$ and arc set $\{(i, j), i, j \in C, i \neq j\}$.

Parameters:

n : number of available vehicles;

s_l : total available supply at depot l;

c_k : the maximum load of vehicle k;

d_i : the latest arrival time required by demand node i, or the expected deadline for node i;

τ_{ijk} : the estimated time to traverse arc (i, j) for vehicle k;it is set to ∞ for nonexistent links;

ζ_i : amount of commodity needed at node i;

Decision Variables:

X_{ijk} : a binary flow variable, equal to 1 if (i, j) is traversed by vehicle k and 0 otherwise;

Y_{ik} : delivery by vehicle k to the demand node i, integer value is assumed;

U_i : amount of unsatisfied demand at node i;

T_{ik} : time at which vehicle k arriving at node i, the unload time is negligible; and

δ_{ik} : delay time happened when vehicle k sends supply to node i.

   If k arrives $i$later than d_i, then $\delta_{ik} > 0$.

Vehicle scheduling should be made so as to minimize the next two objectives. The first one is maximum unsatisfied rate among all demand points. This objective tries to create fairness among all demand points.

$$max\{\frac{U_i}{\zeta_i}, i \in D\} \tag{1}$$

The second one is total weighted time delay. Here, δ_{ik} is the time delay and Y_{ik} is the amount of supply arriving at node i. Total weighted time delay is the product of these two variables. This objective forces supply to arrive before due date.

$$\sum_{i \in D, k \in K} Y_{ik}\delta_{ik} \tag{2}$$

The vehicle routing model (VRM) for emergency supply allocation and transportation with multi-suppliers is formulated as follows:

$$\textbf{min } z_1 = max\{\frac{U_i}{\zeta_i}, i \in D\} \tag{3}$$
$$\textbf{min } z_2 = \sum_{i \in D, k \in K} Y_{ik}\delta_{ik} \tag{4}$$

subject to

$$\sum_{l \in L}\sum_{k \in K(l)}\sum_{j \in D} X_{ljk} \leq n \tag{5}$$

$$\sum_{j \in C} X_{ijk} = \sum_{j \in C} X_{jik} \leq 1 \quad (\forall i \in C, k \in K) \tag{6}$$

$$\sum_{i \in S}\sum_{j \in C \backslash S} X_{ijk} \geq 1 \quad (\forall S \subseteq D, k \in K) \tag{7}$$

$$\sum_{l \in L}\sum_{j \in C} X_{ljk} \leq 1 \quad (\forall k \in K) \tag{8}$$

$$\sum_{k \in K(l)} T_{lk} = 0 \quad (\forall l \in L) \tag{9}$$

$$0 \leq T_{ik} + \tau_{ijk} - T_{jk} \leq (1 - X_{ijk})M \quad (\forall i \in C, j \in D, k \in K) \tag{10}$$

$$0 \leq T_{ik} - \delta_{ik} \leq d_i \sum_{j \in C} X_{ijk} \quad (\forall i \in D, k \in K) \tag{11}$$

$$\delta_{ik} \leq M \sum_{j \in C} X_{ijk} \quad (\forall i \in D, k \in K) \tag{12}$$

$$s_l - \sum_{k \in K(l)} \sum_{i \in D} Y_{ik} \geq 0 \quad (\forall l \in L) \tag{13}$$

$$\sum_{i \in D} Y_{ik} \leq c_k \quad (\forall k \in K) \tag{14}$$

$$Y_{ik} \leq c_k \sum_{j \in D} X_{ijk} \quad (\forall i \in D, k \in K) \tag{15}$$

$$\sum_{k \in K} Y_{ik} + U_i - \zeta_i \geq 0 \quad (\forall i \in D) \tag{16}$$

$$X_{ijk} = \{0, 1\}; T_{ik} \geq 0; Y_{ik} \geq 0; U_i \geq 0; \delta_{ik} \geq 0; \tag{17}$$

The objective of the model is to minimize maximum unsatisfied rate among all demand points and the total weighted time delay. Constraint set (5) specifies that the number of vehicles to service must not exceed the available fleet size. Constraint (6) indicates that each vehicle visits one demand point at most once and the vehicle must leave the demand node without staying there. Constraints (7) are the subtour elimination constraints. A vehicle cannot go to another depot according to constraint (8). This feasible route constraint allow split delivery. Constraints (9)-(12) are time-window constraints that guarantee schedule feasibility with respect to time considerations. Once a vehicle arrives at a demand point later than the required deadline, a penalty $\delta_{ik} \geq 0$ is observed. (13)-(16) state the construction on the commodity flows, while constraint (17) specifies the binary and integer variables.

3. Model analysis

We will prove the above (VRM) problem is NP-hard by showing that the traveling salesman problem(TSP) is a special case of the VRM.

Theorem 1. *The VRM is NP-hard even if there is only one depot with one vehicle.*

Proof. We will construct a special case of VRM, which is a TSP. First, assume that there is only one depot and m demand nodes. Each demand node needs one unit of medical supply with deadline 0. Also assume there is only one vehicle with capacity m at the depot to deliver all m units to the demand nodes. The object is to minimize the total time delay. Under these assumptions the VRM becomes a TSP.

The VRM is a multi-objective model and there are two objectives: maximum unsatisfied rate and total weighted time delay. If we ignore the second objective, WRM can be solved in polynomial time. At first, let's consider the following model:

$$\mathbf{min} \quad max\{\frac{U_i}{\zeta_i}, i \in D\} \tag{18}$$

subject to

$$\sum_{i \in D} (\zeta_i - U_i) \le \sum_{l \in L} s_l \qquad (19)$$

$$U_i \ge 0 \qquad (20)$$

This is a supply allocation model (SAM). By using a simple technique, the SAM can be transformed to the following model:

$$\min \quad \eta \qquad (21)$$

subject to

$$\frac{U_i}{\zeta_i} \le \eta, \quad (\forall i \in D) \qquad (22)$$

$$\sum_{i \in D} (\zeta_i - U_i) \le \sum_{l \in L} s_l \qquad (23)$$

$$U_i \ge 0, \eta \ge 0 \qquad (24)$$

Obviously, this is a linear programming model. It means VRM without the second objective can be easily solved within constraints (5-17).

In many situations, the data such as U_i, ζ_i, s_l in model SAM are integers. Then the supply allocation model with integer constraints (SAMI) is as follows.

$$\min \quad \eta \qquad (25)$$

subject to

$$\frac{U_i}{\zeta_i} \le \eta, \quad (\forall i \in D) \qquad (26)$$

$$\sum_{i \in D} (\zeta_i - U_i) \le \sum_{l \in L} s_l \qquad (27)$$

$$U_i \ge 0 \text{ and integer}, \eta \ge 0 \qquad (28)$$

Next, we will propose an LP-rounding algorithm for the above model and show that this algorithm can find the optimal solution in polynomial time.

Algorithm LPrA LP-ROUNDING ALGORITHM FOR SAMI

---

1. obtain the LP-relaxation of SAMI by deleting all integer constraints.
2. solve LP-relaxation, and get fractional optimal solution (U_i^*, η^*).
3. **for** each U_i^*
4. $U_i = \lceil U_i^* \rceil$
5. **endfor**
6. $a = \sum_{l \in L} s_l - \sum_{i \in D} (\zeta_i - U_i)$
7. **while** $a > 0$
8.   **for** each $i \in D$

9. $\qquad \eta_i = \frac{U_i}{\zeta_i}$

10. $\quad$ **endfor**

11. $\quad \eta = \max_{i \in D}\{\eta_i\}$

12. $\quad$ choose $k \in \{i | \eta_i = \eta\}$

13. $\quad U_k = U_k - 1$

14. $\quad a = a - 1$

15. **endwhile**

16. $\eta = \max_{i \in D}\{\eta_i = \frac{U_i}{\zeta_i}\}$

17. $Q = \{i | \eta_i > \eta^*, \frac{U_i + 1}{\zeta_i} < \eta\}$

18. **while** $Q \neq \Phi$

19. choose j such that $\eta_j = \eta$

20. choose k such that $\frac{U_k + 1}{\zeta_k} = \max_{i \in Q}\{\frac{U_i + 1}{\zeta_i}\}$

21. $U_k = U_k + 1$

22. $U_j = U_j - 1$

23. $\eta = \max_{i \in D}\{\eta_i = \frac{U_i}{\zeta_i}\}$

24. $Q = \{i | \eta_i > \eta^*, \frac{U_i + 1}{\zeta_i} < \eta\}$

25. **endwhile**

26. Output the integer solution U_i and unsatisfied rate η.

Theorem 2. *The algorithm LPrA can find the optimal solution for SAMI in $O(n^2)$ time.*

Proof. First, we will show the algorithm LPrA can stop within $O(n^2)$. According to the definition of a, step 8-14 runs at most n times for $a < n$. Step 11 runs at most n times. Then the total running time from step 7 to 15 is at most n^2. Simultaneously, step 18-25 runs also at most n^2.

Second, we will prove the output solution is optimal by contradiction. Let $\{U_1, U_2, \ldots, U_n\}$ and η be the output solution of the algorithm LPrA. Then for $1 \leq i \leq n$, $\eta_i = \frac{U_i}{\zeta_i} \leq \eta$. Without loss generality, suppose $\eta_k = \eta$. Now, suppose $\{U_1^*, U_2^*, \ldots, U_n^*\}$ and η^* are the optimal solution of SAMI problem, which satisfies $\eta^* < \eta$. Let $\eta_j^* = \eta^*$ without loss of generality. We have $\eta_k = \eta > \eta^* \geq \eta_k^*$, then $U_k > U_k^*$. On the other hand, $\sum U_k = \sum U_k^*$, then there must exit l such that $U_l < U_l^*$. So $\eta_l < \eta_l^* \leq \eta^* < \eta = \eta_k$ holds. Then, $\frac{U_l + 1}{\zeta_l} \leq \eta_l^* < \eta$, $l \in Q$ according to step 17 in algorithm LPrA, a contradiction.

Then, the complexity of VRM is totally up to the second objective, while this model for the second objective is a vehicle routing problem. In the next section, we will propose a local search algorithm.

4. Local search algorithm

Define a supply capability for each depot:

$$M_l = \min\{s_l, \sum_{k \in K(l)} c_k\} \quad (\forall l \in L)$$

The total supply capacity is:
$$M = \sum_{l \in L} M_l$$

Let $\zeta = \sum_{i \in D} \zeta_i$, then we have the next three situations.

1. When $\zeta - M > 0$, this represents that there is not sufficient capacity to deliver all the commodity to the demanding nodes.
2. When $\zeta - M = 0$, there is a balanced capacity for supply and demand.
3. When $\zeta - M < 0$, there is still surplus capacity.

Let $\theta = max\{\zeta - M, 0\}$. When $\theta = 0$, the demand in each node is satisfied.

Before presenting our algorithm, we assume that $|K| < |D|$, i.e. the number of vehicles is less than the number of demand nodes, otherwise the problem will be trivial. We also assume that the route distances (or travel times) follow the triangle inequality, i.e. the direct distance or travel time between any two nodes is less than that through a third node.

The local search algorithm can be divided into two stages: medical allocation and vehicle routing. First, the amount of medical supplies allocated to each node are determined by using algorithm LPrA. Then using a greedy algorithm the vehicles are scheduled. The details of the algorithm are as follows.

Local search algorithm

Step 1: Obtain the amount of supply allocated to each node by using algorithm LPrA, ζ_i'.

Step 2: Let $P = \{p_k\}$ specify the current position of vehicle $k \in K$. Let $U_i = \zeta_i', i \in D$ and let $p_k = l, k \in K(l)$ to specify the vehicles departing from depot l, $(\forall l \in L)$. $T_{lk} = 0$, $\theta = max\{\sum_{i \in D} \zeta_i' - M, 0\}$.

Step 3: Let $Q = \{i | U_i > 0\}$, $K = \{k | c_k > 0\}$.

For all $i \in Q, j \in K$, compute
$$\delta_{ij} = max\{0, T_{p_j} + \tau_{p_j ij} - d_i\}$$
$$Y_{ij} = min\{c_j, U_i\}$$

Find
$$(q, k) \in \{(i, j) | Y_{ij}\delta_{ij} = min_{i \in Q, j \in K}\{Y_{ij}\delta_{ij}\}\}$$

Update
$$X_{p_k qk} := 1$$
$$p_k := q$$
$$Y_{p_k k} := min\{U_{p_k}, c_k\}$$
$$c_k := c_k - Y_{p_k k}$$
$$U_{p_k} := U_{p_k} - Y_{p_k k}$$

Step 4: If $\sum_{i \in D} U_i = \theta$, then $X_{p_k lk} = 1, Y_{lk} = c_k, \forall k \in K(l), l \in L$, stop; otherwise go to Step 3.

5. Simulation

Based on the above model and algorithm, we simulate an emergency situation when a pandemic disease (e.g. SARS) happens in Beijing, China and a certain quantity of medication need to be delivered from the airport and Beijing Emergency Medical Center(EMC) to major downtown hospitals as soon as possible. Name of 16 hospitals and their locations are shown in Table 1 and Figure 1. The distance are shown in table 2. Suppose we have a fleet of 5 identical

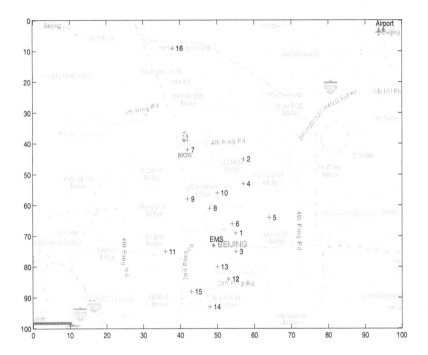

Fig. 1. Hospitals, Airport and EMC in Beijing

trucks in the airport and 3 at the EMC to do the delivery. The capacity of each truck is 25. The average speed of each truck is 40 kilometers per hour. According to the distance matrix, we can obtain the travel time between each pair of nodes. All time units will be represented in minutes. All case study settings are solved on a Windows XP-based Pentium(R) 4 CPU 2.93CHz personal computer using MATLAB 7.0 and its Optimization Toolbox.

There are 125 units of medication at the airport and 75 units at the EMC that need to be sent to the 16 hospitals. The demand and deadline in each hospital are generated randomly from a uniform distribution.

In the first case study, the deadlines are fixed, and we simulate the algorithm 30 times with different demand randomly generated from a uniform distribution. Deadlines are shown in Table 3. Demand in each hospital is generated randomly between 1 to 25 from a uniform

label	hospital
1	Peking Union Medical College Hospital (PUMCH)
2	China-Japan Friendship Hospital (CJFH)
3	Beijing Tongren Hospital (BTrH)
4	Beijing Ditan Hospital (BDH)
5	Beijing Chaoyang Hospital (BCH)
6	Beijing Obstetrics and Gynecology Hospital (BOGH)
7	Peking University Third Hospital (PUTH)
8	Peking University First Hospital (PUFH)
9	Peking University People Hospital (PUPH)
10	Beijing Ji Shui Tan Hospital (BJSYH)
11	Beijing Shijitan Hospital (BSH)
12	Beijing Tiantan Hospital (BTtH)
13	Beijing Friendship Hospital (BFH)
14	China Rehabilitation Research Center (CRRC)
15	Beijing Youan Hospital (BYH)
16	Beijing Hui Long Guan Hospital(BHLGH)

Table 1. Hospitals and their labels

	1	2	3	4	5	6	7	8	9	10	11	12	13	14	15	16	Airport
PUMCT	0	8	1.2	10.6	5	2	12.8	5.7	8.4	7.5	9.1	6.3	5.4	10.4	10.1	21.94	26.36
CJFH	8	0	8.8	4.4	7.8	8.2	7.7	8.8	9.8	8	16.6	13.8	13	21.6	18.4	16.63	22.67
BTrH	1.2	8.8	0	12	5.8	3	13.8	6.6	9.5	8.4	9.4	5	4.4	9	9.5	24.26	27.51
BDH	10.6	4.4	12	0	7.1	4.4	7.8	5.2	6.2	4.3	13	12.5	11.2	20	14.9	17.15	23.91
BCH	5	7.8	5.8	7.1	0	5.1	14.4	7	9.8	8.8	13.8	11.2	10.4	15.3	16.1	19.32	23.73
BOGH	2	8.2	3	4.4	5.1	0	11	3.7	6.5	5.4	9.2	6	5.3	10.8	11.2	21.21	29.14
PUTH	12.8	7.7	13.8	7.8	14.4	11	0	7.5	5.8	6	12.6	17	14.3	17.8	16	11.83	27.72
PUFH	5.7	8.8	6.6	5.2	7	3.7	7.5	0	2.8	1.7	8.5	8	7.6	11.1	9.34	17.73	27.73
PUPH	8.4	9.8	9.5	6.2	9.8	6.5	5.8	2.8	0	3.4	6.8	11.6	9	13.6	8.77	15.66	28.94
BJSYH	7.5	8	8.4	4.3	8.8	5.4	6	1.7	3.4	0	9.4	8.2	8.8	16	9.94	16.49	26.81
BSH	9.1	16.6	9.4	13	13.8	9.2	12.6	8.5	6.8	9.4	0	12	8.7	13.1	7.2	24.65	35.79
BTtH	6.3	13.8	5	12.5	11.2	6	17	8	11.6	8.2	12	0	1.8	4.9	4.6	26.86	32.63
BFH	5.4	13	4.4	11.2	10.4	5.3	14.3	7.6	9	8.8	8.7	1.8	0	6.5	4.88	24	32.12
CRRC	10.4	21.6	9	20	15.3	10.8	17.8	11.1	13.6	16	13.1	4.9	6.5	0	4.8	27.58	37.24
BYH	10.1	18.4	9.5	14.9	16.1	11.2	16	9.34	8.77	9.94	7.2	4.6	4.88	4.8	0	24.56	37.29
BHLGH	21.94	16.63	24.26	17.15	19.32	21.21	11.83	17.73	15.66	16.49	24.65	26.86	24	27.58	24.56	0	32.34
Airport	26.36	22.67	27.51	23.91	23.73	29.14	27.72	27.73	28.94	26.81	35.79	32.63	32.12	37.24	37.29	32.34	0
EMC	4.74	13.6	3.53	9.71	10.41	4.57	12.32	5.06	5.49	5.74	5.69	4.26	2.61	7.49	6.67	34.35	33.66

Table 2. Distance between hospitals, Airport and EMC (Kilometer)

distribution. In the second case study, demand is fixed, then we simulate the algorithm 30 times with different deadline randomly generated. Demand is shown in Table 5. Deadline in each hospital is generated randomly between 40 to 90 minute from a uniform distribution.

hospital	1	2	3	4	5	6	7	8	9	10	11	12	13	14	15	16
deadline	43	44	73	50	83	49	49	90	62	58	56	59	60	70	46	42

Table 3. Fixed deadline of each hospital(minutes)

Problem No.	total supply	total unsatisfied demand	total weighted time delay	maximum unsatisfied rate
1	200	7	1370	0.0667
2	200	6	433	0.0526
3	200	14	193	0.1111
4	200	0	260	0
5	200	61	441	0.2500
6	200	21	949	0.1304
7	200	20	260	0.1200
8	200	0	452	0
9	200	0	48	0
10	200	0	499	0
11	200	38	1443	0.2000
12	200	0	599	0
13	200	3	376	0.0476
14	200	38	620	0.1875
15	200	16	1499	0.1176
16	200	15	99	0.1000
17	200	51	356	0.2353
18	200	3	228	0.0500
19	200	0	806	0
20	200	52	16	0.2381
21	200	31	198	0.1667
22	200	0	161	0
23	200	0	416	0
24	200	2	1120	0.0417
25	200	25	445	0.1500
26	200	0	585	0
27	200	0	54	0
28	200	4	627	0.0455
29	200	0	18	0
30	200	72	350	0.2857
average value	200	15.9667	497.3667	0.0866

Table 4. Simulation results

hospital	1	2	3	4	5	6	7	8	9	10	11	12	13	14	15	16
demand	13	25	10	14	5	13	11	17	17	24	5	3	15	25	1	22

Table 5. Fixed demand of each hospital

Problem No.	total supply	total unsatisfied demand	total weighted time delay	maximum unsatisfied rate
1	200	20	0	0.12
2	200	20	660	0.12
3	200	20	321	0.12
4	200	20	422	0.12
5	200	20	880	0.12
6	200	20	63	0.12
7	200	20	338	0.12
8	200	20	0	0.12
9	200	20	346	0.12
10	200	20	204	0.12
11	200	20	272	0.12
12	200	20	28	0.12
13	200	20	45	0.12
14	200	20	21	0.12
15	200	20	201	0.12
16	200	20	223	0.12
17	200	20	87	0.12
18	200	20	33	0.12
19	200	20	1261	0.12
20	200	20	46	0.12
21	200	20	233	0.12
22	200	20	237	0.12
23	200	20	109	0.12
24	200	20	6	0.12
25	200	20	312	0.12
26	200	20	432	0.12
27	200	20	40	0.12
28	200	20	188	0.12
29	200	20	725	0.12
30	200	20	60	0.12
average value	200	20	259.7667	0.12

Table 6. Simulation results

From these two computational cases, the following observations can be made.

- For fixed deadline cases, the average maximum unsatisfied rate is 0.0866. Medical supplies are allocated equitably. The same conclusion can be made for fixed demand.
- For each instance, the local search algorithm can present an efficient vehicle routing schedule.
- The algorithm runs fast.

We specify another model where fairness is not considered. That means the first objective in VRM is ignored, and we specify this model as VRM'. The aim of proposing this model is to compare with VRM. Firstly, given three problems which information is shown in Table 7. Table

Problem No.		
1	deadline	77,61,60,66,49,67,73,41,82,81,75,64,45,82,50,63
	demand	1,8,22,21,9,23,12,15,16,17,16,18,13,18,13,16
	total demand	238
	total supply	200
2	deadline	89,82,56,70,47,53,81,74,41,69,63,86,55,44,64,90
	demand	24,15,17,20,3,1,14,1,12,5,20,16,1,23,20,23
	total demand	215
	total supply	200
3	deadline	78,60,57,66,69,79,79,65,81,64,51,69,74,74,88,79
	demand	19,22,25,13,16,20,12,14,5,4,6,3,4,12,20,8
	total demand	203
	total supply	200

Table 7. Problem information

8 shows the comparison between the solutions obtained from the above two models. When the total unsatisfied demand is big as that of problem 1, the unsatisfied rate, obtained from VRMạf, may be worse even though its total weighted time delay is smaller than that of VRM. While total unsatisfied demand is small, the VRM can present better solution than VRM', such as problem 3. A sample vehicle routing scheme when information is confirmed. Each type of line is corresponding to one vehicle route. For example, one vehicle drives from airport to hospital 8, 13, 6, 9 according to the red line. With different data input we have simulated cases

Problem No.	maximum unsatisfied demand		total weighted time delay		maximum unsatisfied rate	
	VRM	VRM'	VRM	VRM'	VRM	VRM'
1	38	38	325	435	0.1875	1
2	15	15	343	304	0.1	0.625
3	3	3	46	46	0.05	0.15

Table 8. Comparison between the above two models

of severe supply shortage, tight deadline, large fleet, and large number of randomly generated demand nodes. All these results show that the polynomial time algorithm is very efficient and can be a very useful tool in routing vehicles during a large-scale emergency scenario.

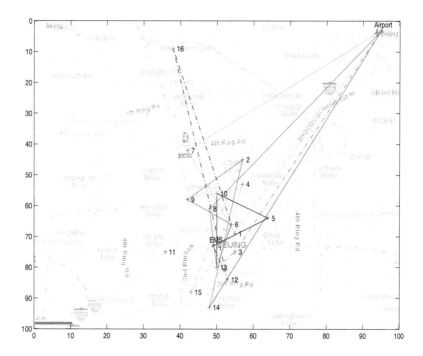

Fig. 2. Example for vehicles routing scheme

6. Conclusion

In this paper, we consider the vehicle routing problem under an emergency situation. A multi-objective model is formulated. Supplies may arrive with time delay, and the first objective is to minimize the total delay. We also consider fairness among demand nodes with respect to their unsatisfied rates. A new model and local search algorithm are presented. Simulation results show that the algorithm can be very useful for emergency responder to effectively use the available vehicles in case of emergencies.

For future work, we are going to design new algorithms for this model by using some other technique, such as heuristic algorithms. We will also try to formulate new models, when information about the demand and deadline are uncertain.

7. References

Astrid, S. K. & David, P. M. (2003). Stochastic vehicle routing with random travel times. *Transportation Science*, Vol. 37(1):69-82.

Altay, N. & Green, III W. G. (2006). OR/MS research in disaster operations management. *European Journal of Operations Research*, Vol. 175(1):475-493.

Barbarosoglu, G. & Arda, Y. (2004). A two-stage stochastic programming framework for transportation planning in disaster response. *Journal of the Operational Research Society*, Vol. 55:43-53.

Campbell, A. M. Vandenbussche, D. & Hermann, W. (2008). Routing for relief efforts. *Transportation Science*, Vol. 42(2):127-45.

Dantzig, G. B., Ramser, J.H. (1959). The truck dispatching problem. *Management Science*, Vol. 6(1):80-91.

Enrico, A. & Maria, G. S. (2002). The periodic vehicle routing problem with intermediate facilities. *European Journal of Operations Research*, Vol. 137:233-247.

Fiedrich, A., Gehbauer, F. & Rickers U. (2000). Optimized resource allocation for emergency response after earthquake disasters. *Safety Science*, Vol. 35:41-57.

Haghani, A. & Oh, S. (1996). Formulation and solution of a multi-commodity, multi-modal network flow model for disaster relief operations. *Transportation Research A*, Vol. 30(3):231-250.

Laporte, G. (1992). The vehicle routing problem: An overview of exact and approximate algorithms. *European Journal of Operations Research*, Vol. 59:345-358.

Larson, R. C., Metzger, M. D. & Cahn, M.F. (2006). Responding to emergencies: Lessons learned and the need for analysis. *Interfaces*, Vol. 36(6):486-501.

Liu, D., Han, J. & Zhu, J. (2007). Vehicle Routing For Medical Supplies in large-scale Emergencies. *Lecture Notes in Operations Research*, Vol. 8:412-419.

Lorini, S., Potvin, J., & Zufferey, N. (2011). Online vehicle routing and scheduling with dynamic travel times. *Computers & Operations Research*, Vol. 38:1086-1090.

Ngueveu, S. U., Prins, C. & Calvo, R.W. (2010). An effective memetic algorithm for the cumulative capacitated vehicle routing problem. *Computers and Operations Research*, Vol. 37(11):1877-1885.

Rathi, A. H., Church, R. & Solanki R. (1993). Allocating resources to support a mutlicommodity flow with time windows. *Logistics and Transportation Review*, Vol. 28(2):167-188.

Roberto, B., Enrico, B., Aristide, M. & Andrea, V. (2011). An exact algorithm for the period routing problem. *Operations Research*, Vol. 59(1):228-241.

Renaud, J., Laporte, G. & Boctor, F.F. (1996). A tabu search heuristic for the multidepot vehicle routing problem. *Computers and Operations Research*, Vol. 23:229-235.

Sheu, J. (2007). Challenges of Emergency Logistics Management. *Transportation Research Part E*, Vol. 43(6):655-659.

Shen, Z., Dessouky, M. & Ordonez, F. (2007). Stochastic vehicle routing problem for large-scale emergencies. *http://illposed.usc.edu/ fordon/docs/routing4LSE.pdf, November.*

Toth, P. & Vigo, D. (2002). The Vehicle Routing Problem. *SIAM Monographs on Discrete Mathematics and Applications*, SIAM Publishing.

Yi, W. & Ozdamar, L. (2004). Fuzzy modeling for coordinating logistics in emergencies. *International Scientific Journal of Methods and Models of Complexity-Special Issue on Societal Problems in Turkey*, Vol. 7(1):2-24

Permissions

The contributors of this book come from diverse backgrounds, making this book a truly international effort. This book will bring forth new frontiers with its revolutionizing research information and detailed analysis of the nascent developments around the world.

We would like to thank Prof. Burak Eksioglu, for lending his expertise to make the book truly unique. He has played a crucial role in the development of this book. Without his invaluable contribution this book wouldn't have been possible. He has made vital efforts to compile up to date information on the varied aspects of this subject to make this book a valuable addition to the collection of many professionals and students.

This book was conceptualized with the vision of imparting up-to-date information and advanced data in this field. To ensure the same, a matchless editorial board was set up. Every individual on the board went through rigorous rounds of assessment to prove their worth. After which they invested a large part of their time researching and compiling the most relevant data for our readers. Conferences and sessions were held from time to time between the editorial board and the contributing authors to present the data in the most comprehensible form. The editorial team has worked tirelessly to provide valuable and valid information to help people across the globe.

Every chapter published in this book has been scrutinized by our experts. Their significance has been extensively debated. The topics covered herein carry significant findings which will fuel the growth of the discipline. They may even be implemented as practical applications or may be referred to as a beginning point for another development. Chapters in this book were first published by InTech; hereby published with permission under the Creative Commons Attribution License or equivalent.

The editorial board has been involved in producing this book since its inception. They have spent rigorous hours researching and exploring the diverse topics which have resulted in the successful publishing of this book. They have passed on their knowledge of decades through this book. To expedite this challenging task, the publisher supported the team at every step. A small team of assistant editors was also appointed to further simplify the editing procedure and attain best results for the readers.

Our editorial team has been hand-picked from every corner of the world. Their multi-ethnicity adds dynamic inputs to the discussions which result in innovative outcomes. These outcomes are then further discussed with the researchers and contributors who give their valuable feedback and opinion regarding the same. The feedback is then collaborated with the researches and they are edited in a comprehensive manner to aid the understanding of the subject.

Apart from the editorial board, the designing team has also invested a significant amount of their time in understanding the subject and creating the most relevant covers. They scrutinized every image to scout for the most suitable representation of the subject and create an appropriate cover for the book.

The publishing team has been involved in this book since its early stages. They were actively engaged in every process, be it collecting the data, connecting with the contributors or procuring relevant information. The team has been an ardent support to the editorial, designing and production team. Their endless efforts to recruit the best for this project, has resulted in the accomplishment of this book. They are a veteran in the field of academics and their pool of knowledge is as vast as their experience in printing. Their expertise and guidance has proved useful at every step. Their uncompromising quality standards have made this book an exceptional effort. Their encouragement from time to time has been an inspiration for everyone.

The publisher and the editorial board hope that this book will prove to be a valuable piece of knowledge for researchers, students, practitioners and scholars across the globe.

List of Contributors

Jessie J. Walker
University of Arkansas at Pine Bluff/Computer Science Unit, USA

David Hutton
Deputy Director, United Nations Relief and Works Agency

Gabriele Mencagli and Marco Vanneschi
Department of Computer Science, University of Pisa, L. Bruno Pontecorvo, Pisa, Italy

Yue Liu
University of Wisconsin-Milwaukee, USA

Jie Yu
Shandong University, China

Jianming Zhu
College of Engineering, Graduate University of Chinese Academy of Sciences, China